AF553810

TEXTBOOK OF DAIRY CHEMISTRY

TEXTBOOK OF DAIRY CHEMISTRY

By

Dr. Arvind N. Shukla
School of Studies of Zoology & Biotechnology
Vikram University
Ujjain (India)

DISCOVERY PUBLISHING HOUSE PVT. LTD.
NEW DELHI-110 002

Published by:
Namit Wasan

DISCOVERY PUBLISHING HOUSE PVT. LTD.
4383/4B, Ansari Road, Darya Ganj
New Delhi-110 002 (India)
Phone : +91-11-23279245; 23253475; 43596065
E-mail : discoverybooksindia@gmail.com
discoverypublishinghouse@gmail.com
namitwasan9@gmail.com
web : www.discoverypublishinggroup.com

***First Edition:* 2010**

***Reprinted:* 2024**

ISBN: 978-81-8356-571-4

Textbook of Dairy Chemistry

Printed at:
Infinity Imaging Systems
Delhi

PREFACE

Milk is as ancient as man-kind itself, as it is the substance created to feed the mammalian infant. All species of mammals, from man to whales, produce milk for this purpose. Many centuries ago, perhaps as early as 6000-8000 BC, ancient man learned to domesticate species of animals for the provision of milk to be consumed by them. These included cows (genus Bos), buffaloes, sheep, goats, and camels, all of which are still used in various parts of the world for the production of milk for human consumption.

Fermented products such as cheeses were discovered by accident, but their history has also been documented for many centuries, as has the production of concentrated milks, butter, and even ice-cream.

Technological advances have only come about very recently in the history of milk consumption, and our generations will be the ones credited for having turned milk processing from an art to a science. The availability and distribution of milk and milk products today in the modern world is a blend of the centuries old knowledge of traditional milk products with the application of modern science and technology.

The role of milk in the traditional diet has varied greatly in different regions of the world. The tropical countries have not been traditional milk consumers, whereas the more northern regions of the world, Europe (especially Scandinavia) and North America, have traditionally consumed far more milk and milk products in their diet. In tropical countries where high temperatures and lack of refrigeration has led to the inability to produce and store fresh milk, milk has traditionally been preserved through means other than refrigeration, including immediate consumption of warm milk after milking, by boiling milk, or by conversion into more stable products such as fermented milks.

—Author

CONTENTS

1

INTRODUCTION

Dairy chemistry is the study of milk and milk-derived food products from a food science perspective. It focuses on the biological, chemical, physical, and microbiological aspects of milk itself, and on the technological (processing) aspects of the transformation of milk into its various consumer products, including beverages, fermented products, concentrated and dried products, butter and ice cream.

Milk is as ancient as man-kind itself, as it is the substance created to feed the mammalian infant. All species of mammals, from man to whales, produce milk for this purpose. Many centuries ago, perhaps as early as 6000-8000 BC, ancient man learned to domesticate species of animals for the provision of milk to be consumed by them. These included cows (genus Bos), buffaloes, sheep, goats, and camels, all of which are still used in various parts of the world for the production of milk for human consumption.

Fermented products such as cheeses were discovered by accident, but their history has also been documented for many centuries, as has the production of concentrated milks, butter, and even ice-cream.

Technological advances have only come about very recently in the history of milk consumption, and our generations will be the ones credited for having turned milk processing from an art to a science. The availability and distribution of milk and milk products today in the modern world is a blend of the centuries old knowledge of traditional milk products with the application of modern science and technology.

The role of milk in the traditional diet has varied greatly in different regions of the world. The tropical countries have not been traditional milk consumers, whereas the more northern regions of the world, Europe (especially Scandinavia) and North America, have traditionally consumed far more milk and milk products in their diet. In tropical countries where high temperatures and lack of refrigeration has led to the inability to produce and store fresh milk, milk has traditionally been preserved through means other than refrigeration, including immediate consumption of warm milk after milking, by boiling milk, or by conversion into more stable products such as fermented milks.

WORLD-WIDE MILK CONSUMPTION AND PRODUCTION

The total milk consumption (as fluid milk and processed products) per person varies widely from highs in Europe and North America to lows in Asia. However, as the various regions of the world become more integrated through travel and migration, these trends are changing, a factor which needs to be considered by product developers and marketers of milk and milk products in various countries of the world.

Even within regions such as Europe, the custom of milk consumption has varied greatly. Consider for example the high consumption of fluid milk in countries such as Finland, Norway and Sweden compared to France and Italy where cheeses have tended to dominate milk consumption. When you also consider the climates of these regions, it would appear that the culture of producing more stable products (cheese) in hotter climates as a means of preservation is evident.

2

Milk Production and Biosynthesis

INTRODUCTION

Milk is the source of nutrients and immunological protection for the young cow. The gestation period for the female cow is nine months. Shortly before calving, milk is secreted into the udder in preparation for the new born. At parturition, fluid from the mammary gland known as colostrum is secreted. This yellowish coloured, salty liquid has a very high serum protein content and provides antibodies to help protect the newborn until its own immune system is established. Within 72 hours, the composition of colostrum returns to that of fresh milk, allowing to be used in the food supply.

The period of lactation, or milk production, then continues for an average of 305 days, producing as much as 9000 or more kg of milk. This is quite a large amount considering the calf only needs about 1000 kg for growth.

Within the lactation, the highest yield is 2-3 months post-parturition, yielding 40-50 l/day. Within the milking lifetime, a cow reaches a peak in production about her third lactation, but can be kept in production for 5-6 lactations if the yield is still good.

About 1-2 months after calving, the cow begins to come into heat again. She is usually inseminated about 3 months after calving so as to come into a yearly calving cycle. Heifers are normally first inseminated at 15 months so she's 2 when the first calf is born. About 60 days before the next calving, the cow is dried off. There is no milking during this stage for two reasons:

- milk has tapered off because of maternal needs of the fetus; and
- udder needs time to prepare for the next milking cycle.

The life of a female cow can be summerized in Table 2.1.

Table 2.1: The life of a female cow

Age	
0	Calf born
15 months	Heifer inseminated for first calf
24 months	First calf born - starts milking
27 months	Inseminated for second calf
34 months	Dried off
36 months	Second calf born - starts milking

Cycle repeats for 5-6 lactations.

CLEANLINESS

The environment of production has a great effect on the quality of milk produced. From the food science perspective, the production of the highest quality milk should be the goal. However, this is sometimes not the greatest concern of those involved in milk production. Hygienic quality assessment tests include sensory tests, dye reduction tests for microbial activity, total bacterial count (standard plate count), sediment, titratable acidity, somatic cell count, antibiotic residues, and added water.

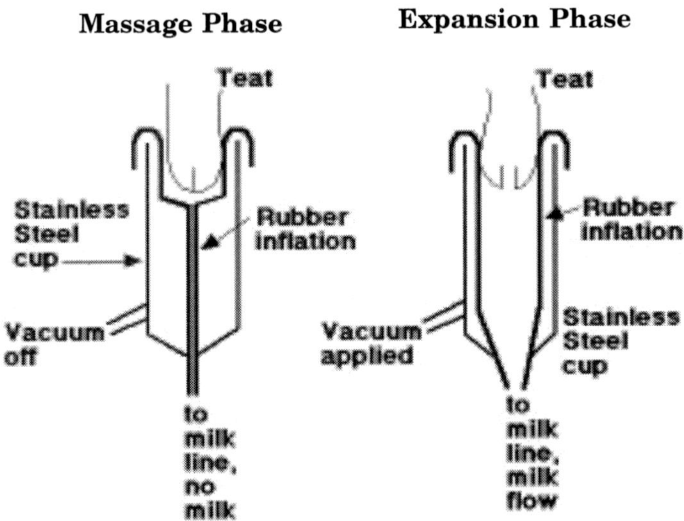

Fig 2.1: Effects of Milk Handling on Quality and Hygiene

The two common dye reduction tests are methylene blue and resazurin. These are both synthetic compounds which accept electrons and change colour as a result of this reduction. As part of natural metabolism, active micro-organisms transfer electrons, and thus rate at which dyes added to milk are reduced is an indication of the level of microbial activity. Methylene blue turns from blue to colorless, while resazurin turns from blue to violet to pink to colourless. The reduction time is inversely correlated to bacterial numbers. However, different species react differently. Mesophilics are favoured over psychrotrophs, but psychrotrophic organisms tend to be more numerous and active in cooled milk.

Temperature

Milk production and distribution in the tropical regions of the world is more challenging due to the requirements for low-temperature for milk stability. Table 2.2 illustrates the numbers of bacteria per millilitre of milk after 24 hours.

Table 2.2: Number of bacteria per ml of milk after 24 hours

5°C	2,600
10°C	11,600
12.7°C	18,800
15.5°C	180,000
20°C	450,000

Traditionally, this has been overcome in tropical countries by stabilizing milk through means other than refrigeration, including immediate consumption of warm milk after milking, by boiling milk, or by conversion into more stable products such as fermented milks.

MASTITIS AND ANTIBIOTICS

Mastitis is a bacterial and yeast infection of the udder. Milk from mastitic cows is termed abnormal. Its SNF, especially lactose, content is decreased, while Na and Cl levels are increased, often giving mastitic milk a salty flavour. The presence of mastitis is also accompanied by increases in bacterial numbers, including the possibility of human pathogens, and by a dramatic increase in somatic cells. These are comprised of leukocytes (white blood cells) and epithelial cells from the udder lining. Increased somatic cell counts are therefore indicative of the presence of mastitis. Once the infection reaches the level known as "clinical' mastitis, pus can be observed in the teat canal just prior to milking, but at sub-clinical levels, the presence of mastitis is not obvious.

Table 2.3

Somatic Cell Count (000's/ml)tations	Daily Milk Yield (kg)	1st Lactation Older Lac
0-17	23.1	29.3
18-34	23.0	28.7
35-70	22.6	28.0
71-140	22.4	27.4

(Contd...)

141-282	22.1	27.0
282-565	21.9	26.3
566-1130	21.4	25.4
1131-2262	20.7	24.6
2263-4525	20.0	23.6
>4526	19.0	22.5

Antibiotics are frequently used to control mastitis in dairy cattle. However, the presence of antibiotic residues in milk is very problematic, for at least three reasons. In the production of fermented milks, antibiotic residues can slow or destroy the growth of the fermentation bacteria. From a human health point of view, some people are allergic to specific antibiotics, and their presence in food consumed can have severe consequences. Also, frequent exposure to low level antibiotics can cause microorganisms to become resistant to them, through mutation, so that they are ineffective when needed to fight a human infection. For these reasons, it is extremely important that milk from cows being treated with antibiotics is withheld from the milk supply.

The withdrawal time after final treatment for various antibiotics is shown below:

Amoxcillin 60 h.

Cloxacillin 48 h.

Erythromicin 36 h.

Novobiocin 72 h.

Penicillin 84 h.

Sulfadimethozine 60 h.

Sulfabromomethozine 96 h.

Sulfaethoxypyridozine 72 h.

ANTI-MICROBIAL SYSTEMS IN RAW MILK

There exists in milk a number of natural anti-microbial defence mechanisms. These include:

(i) *Lysozyme* - an enzyme that hydrolyses glycosidic bonds in gram positive cell walls. However, its effect as a bacteriostatic mechanism in milk is probably negligible.

(ii) *Lactoferrin* - an iron binding protein that sequesters iron from microorganisms, thus taking away one of their growth factors. Its effect as a bacteriostatic mechanism in milk is also probably negligible.

(iii) *Lactoperoxidase* - an enzyme naturally present in raw milk that catalyzes the conversion of hydrogen peroxide to water. When hydrogen peroxide and thiocyanate are added to raw milk, the thiocyanate is oxidized by the enzyme/hydrogen peroxide complex producing bacteriostatic compounds that inhibit Gram negative bacteria, *E. coli*, *Salmonella* spp., and *streptococci*. This technique is being used in many parts of the world, especially where refrigeration for raw milk is not readily available, as a means of increasing the shelf life of raw milk.

Milk Biosynthesis

Milk is synthesized in the mammary gland. Within the mammary gland is the milk producing unit, the alveolus. It contains a single layer of epithelial secretory cells surrounding a central storage area called the lumen, which is connected to a duct system. The secretory cells are, in turn, surrounded by a layer of myoepithelial cells and blood capillaries. The raw materials for milk production are transported via the bloodstream to the secretory cells. It takes 400-800 l of blood to deliver components for 1 l of milk.

Proteins

Building blocks are amino acids in the blood. Casein micelles, or small aggregates thereof, may begin aggregation in Golgi vesicles within the secretory cell.

Lipids

C4-C14 fatty acids are synthesized in the cells. C16 and greater fatty acids are preformed as a result of rumen hydrogenation and are transported directly in the blood.

Lactose

Milk is in osmotic equilibrium with the blood and is controlled by lactose, K, Na, Cl; lactose synthesis regulates the volume of milk secreted. The milk components are synthesized within the cells, mainly by the endoplasmic reticulum (ER) and its attached ribosomes. The energy for the ER is supplied by the mitochondria. The components are then passed along to the Golgi apparatus, which is responsible for their eventual movement out of the cell in the form of vesicles. Both vesicles containing aqueous non-fat components, as well as liquid droplets (synthesized by the ER) must pass through the cytoplasm and the apical plasma membrane to be deposited in the lumen. It is thought that the milk fat globule membrane is comprised of the apical plasma membrane of the secretory cell.

Milking stimuli, such as a sucking calf, a warm wash cloth, the regime of parlour etc., causes the release of a hormone called oxytocin. Oxytocin is relased from the pituitary gland, below the brain, to begin the process of milk let-down. As a result of this hormone stimulation, the muscles begin to compress the alveoli, causing a pressure in the udder known as letdown reflex, and the milk components stored in the lumen are released into the duct system. The milk is forced down into the teat cistern from which it is milked. The let-down reflex fades as the oxytocin is degraded, within 4-7 minutes. It is very difficult to milk after this time.

3

Milk Grading and Defects

INTRODUCTION

The importance of milk grading lies in the fact that dairy products are only as good as the raw materials from which they were made. It is important that dairy personnel have a knowledge of sensory perception and evaluation techniques. The identification of off-flavours and desirable flavours, as well as knowledge of their likely cause, should enable the production of high quality milk, and subsequently, high quality dairy products.

- Milk Grading;
- Sense of Taste;
- Sense of Smell;
- Techniques;
- Milk Defects;
- Lipolyzed;
- Oxidiation;
- Sunlight;
- Cooked;
- Transmitted;

- Microbial; and
- Milk Grading.

An understanding of the principles of sensory evaluation are neccessary for grading. All five primary senses are used in the sensory evaluation of dairy products: sight, taste, smell, touch and sound. The greatest emphasis, however, is placed on taste and smell.

The Sense of Taste

Taste buds, or receptors, are chiefly on the upper surface of the tongue, but may also be present in the cheek and soft palates of young people. These buds, about 900 in number, must make contact with the flavouring agent before a taste sensation occurs. Saliva, of course, is essential in aiding this contact. There are four different types of nerve endings on the tongue which detect the four basic "mouth" flavours—sweet, salt, sour, and bitter. Samples must, therefore, be spread around in the mouth in order to make positive flavour identification. In addition to these basic tastes, the mouth also allows us to get such reactions as coolness, warmth, sweetness, astringency, etc.

The Sense of Smell

We are much more perceptive to the sense of smell than we are to taste. For instance, it is possible for an odouriferous material such as mercaptain to be detected in 20 billion parts of air. The centres of olfaction are located chiefly in the uppermost part of the nasal cavity. To be detectable by smell, a substance must dissolve at body temperature and be soluble in fat solvents.

Note: The sense of both taste and smell may become fatigued during steady use. A good judge does not try to examine more than one sample per minute. Rinsing the mouth with water between samples may help to restore sensitivity.

MILK GRADING TECHNIQUES

Temperature should be between 60-70° F (15.5-21° C) so that any odour present may be detected readily by sniffing the container. Also, we want a temperature rise when taking the sample into the mouth; this serves to volatize any notable constituents. Noting the odour by placing the nose directly over the container immediately after shaking and taking a full "whiff" of air. Any off odour present may be noted.

Need to make sure we have a representative sample; mixing and agitation are important.

Agitation leaves a thin film of milk on the inner surface which tends to evaporate giving off odour if present. During sampling, take a generous sip, roll about the mouth, note flavour sensation, and expectorate. Swallowing milk is a poor practice. Can enhance the after-taste by drawing a breath of fresh air slowly through the mouth and then exhale slowly through the nose. With this practice, even faint odours can be noted. Milk has a flavour defect if it has an odour, a foretaste or an aftertaste, or does not leave the mouth in a clean, sweet, pleasant condition after tasting.

CHARACTERIZATION OF FLAVOUR DEFECTS - ADSA

Lipolytic or Hydrolytic Rancidity

Rancidity arises from the hydrolysis of milkfat by an enzyme called the lipoprotein lipase (LPL). The flavour is due to the short chain fatty acids produced, particularly butyric acid. The LPL can be indigenous or bacterial. It is active at the fat/water interface but is ineffective unless the fat globule membrane is damaged or weakened. This may occur through agitation, and/or foaming, and pumping. For this reason, homogenized milk is subject to rapid lipolysis unless lipase is destroyed by heating first; the enzyme (protein) is denatured at 55-60° C. Therefore, always homogenize milk immediately before or after pasteurization and avoid mixing new and homogenized milk because it leads to rapid rancidity.

Some cows can produce spontaneous lipolysis from reacting to something indigenous to the milk. Late lactation, mastitis, hay and grain ratio diets (more so than fresh forage or silage), and low yielding cows are more suseptible.

Lipolysis can be detected by measuring the acid degree value which determines the presence of free fatty acids. Lipolytic or hydrolytic rancidity is distinct from oxidative rancidity, but frequently in other fat industries, rancid is used to mean oxidative rancidity; in dairy, rancidity means lipolysis.

Characterized: soapy, blue-cheese like aroma, slightly bitter, foul, pronounced aftertaste, does not clear up readily.

Oxidation

Milk fat oxidation is catalysed by copper and certain other metals with oxygen and air. This leads to an autooxidation reaction consisting of initiation, propagation, termination.

RH — R + H initiation - free radical

$R + O_2$ — RO_2 propagation

RO_2 + RH — ROOH + R

R + R — R_2 termination

R + RO_2 — RO_2R

It is usually initiated in the phospholipid of the fat globule membrane. Propagation then occurs in triglycerides, primarily double bonds of unsaturated fatty acids. During propagation, peroxide derivatives of fatty acids accumulate. These undergo further reactions to form carbonyls, of which some, like aldehydes and ketones, have strong flavours. Dry feed, late lactation, added copper or other metals, lack of Vit E (tocopherol) or selenium (natural antioxidates) in the diet all lead to spontaneous oxidation. It can be a real problem especially in winter. Exposure to metals during processing can also contribute.

Characterized: metallic, wet cardboard, oily, tallowy, chalky; mouth usually perceives a puckery or astringent feel.

Sunlight

Often confused with oxidized, this defect is caused by UV-rays from sunlight or flourescent lighting catalyzing oxidation in unprotected milk. Photo-oxidation activates riboflavin which is responsible for catalyzing the conversion of methionine to methanal. It is, therefore, a protein reaction rather than a lipid reaction. However, the end product flavour notes are similar but tends to diminish after storage of several days.

Characterized: burnt-protein or burnt-feathers-like, "medicinal"-like flavour.

Cooked

This defect is a function of the time-temperature of heating and especially the presence of any "burn-on" action of heat on certain proteins, particulary whey proteins. Whey proteins are a source of sulfide bonds which form sulfhydryl groups that contribute to the flavour. The defect is most obvious immediately after heating but dissipates within 1 or 2 days.

Characterized: slightly cooked or nutty-like to scorched or caramelized.

Transmitted Flavours

Cows are particulary bad for transmitting flavours through milk and milk is equally as susceptible to pick-up of off flavours in storage. Feed flavours and green grass can be problems so it is necessary to remove cows from feed 2-4 h before milking. Weeds, garlic/onion, and dandelions can tranfer flavours to the milk and even subsequent products such as butter. Barny flavours can be picked up in the milk if there is poor ventilation and the barn is not properly cleared and cows breathe the air. These flavours are volatile so can be driven off through vacuum de-aeration.

Microbial

There are many flavour defects of dairy products that may be caused by bacteria, yeasts, or moulds. In raw milk the high acid/sour flavour is caused by the growth of lactic acid bacteria which ferment lactose. It is less common today due to change in raw milk microflora. In both raw or processed milk, fruity flavours may arise due to psychrotrophs such as Pseudomonas fragi. Bitter or putrid flavours are caused by psychrotrophic bacteria which produce protease. It is the proteolytic action of protease that usually causes spoilage in milk. Malty flavours are caused by *S.lactis* var. maltigenes and is characterized by a corn flakes type flavour. Although more of a tactile defect, ropy milk is also caused by bacteria, specifically those which produce exopolysaccharides.

Flavour Criticisms	**Intensity of Defect**		
	Slight	**Definite**	**Pronounced**
Astringent	8	7	5
Barny	7	5	3
Bitter	7	5	3
Cooked	9	8	6
Cowy	6	4	1
Feed	9	7	5
Flat	9	8	7
Foreign	5	3	0
Garlic/onion	5	3	1
High acid	3	1	0
Bacterial	5	3	0
Lacks Freshness	7	5	3
Malty	7	5	3
Oxidized	7	5	3
Rancid	7	5	3
Salty	8	6	4
Unclean	7	5	3

4

Dairy Chemistry

COMPOSITION AND STRUCTURE

The role of milk in nature is to nourish and provide immunological protection for the mammalian young. Milk has been a food source for humans since prehistoric times; from human, goat, buffalo, sheep, yak, to the focus of this section - domesticated cow milk (genus Bos). Milk and honey are the only articles of diet whose sole function in nature is food. It is not surprising, therefore, that the nutritional value of milk is high. Milk is also a very complex food with over 100,000 different molecular species found. There are many factors that can affect milk composition such as breed variations (see introduction, cow to cow variations, herd to herd variations - including management and feed considerations, seasonal variations, and geographic variations. With all this in mind, only an approximate composition of milk can be given:

- 87.3% water (range of 85.5% - 88.7%)
- 3.9 % milkfat (range of 2.4% - 5.5%)
- 8.8% solids-not-fat (range of 7.9 - 10.0%)

    (a) protein 3.25% (3/4 casein)

    (b) lactose 4.6%

(c) minerals 0.65% - Ca, P, citrate, Mg, K, Na, Zn, Cl, Fe, Cu, sulfate, bicarbonate, many others

(d) acids 0.18% - citrate, formate, acetate, lactate, oxalate

(e) enzymes - peroxidase, catalase, phosphatase, lipase

(f) gases - oxygen, nitrogen

(g) vitamins - A, C, D, thiamine, riboflavin, others

The following terms are used to describe milk fractions:

- Plasma = milk – fat (skim milk)
- Serum = plasma – casein micelles (whey)
- Solids-not-fat (SNF) = proteins, lactose, minerals, acids, enzymes, vitamins
- Total Milk Solids = fat + SNF

Milk Structure

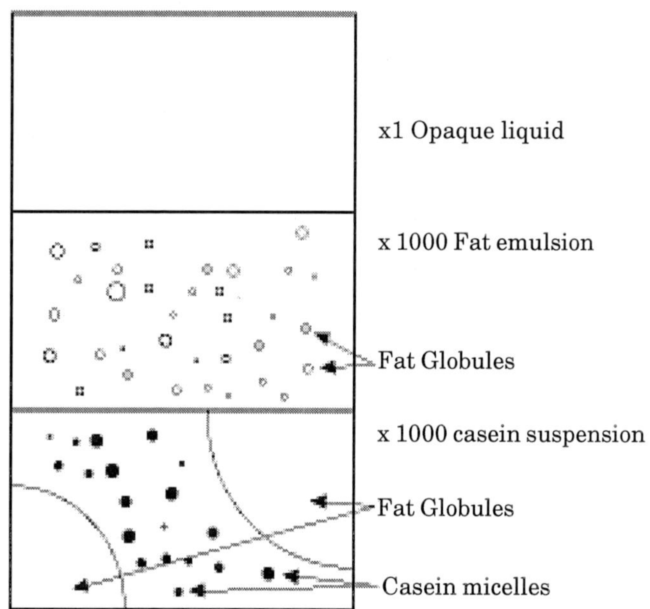

Fig 4.1

Not only is the composition important in determining the properties of milk, but the physical structure must also be examined. Due to its role in nature, milk is in a liquid form. This may seem curious if one takes into consideration the fact that milk has less water than most fruits and vegetables. Milk can be described as:

- an *oil-in-water emulsion* with the fat globules dispersed in the continuous serum phase;
- a *colloid suspension* of casein micelles, globular proteins and lipoprotein partilcles;
- a *solution* of lactose, soluble proteins, minerals, vitamins other components.

Looking at milk under a microscope, at low magnification (5X) a uniform but turbid liquid is observed. At 500X magnification, spherical droplets of fat, known as fat globules, can be seen. At even higher magnification (50,000X), the casein micelles can be observed. The main structural components of milk, fat globules and casein micelles, will be examined in more detail later.

MILK LIPIDS - CHEMICAL PROPERTIES

The fat content of milk is of economic importance because milk is sold on the basis of fat. Milk fatty acids originate either from microbial activity in the rumen, and transported to the secretory cells via the blood and lymph, or from synthesis in the secretory cells. The main milk lipids are a class called triglycerides which are comprised of a glycerol backbone binding up to three different fatty acids. The fatty acids are composed of a hydrocarbon chain and a carboxyl group. The major fatty acids found in milk are:

Long Chain

- C14 - myristic 11%
- C16 - palmitic 26%
- C18 - stearic 10%
- C18:1 - oleic 20%

Short chain (11%)

- C4 - butyric*
- C6 - caproic
- C8 - caprylic
- C10 - capric

* butyric fatty acid is specific for milk fat of ruminant animals and is resposible for the rancid flavour when it is cleaved from glycerol by lipase action.

Saturated fatty acids (no double bonds), such as myristic, palmitic, and stearic make up two thirds of milk fatty acids. Oleic acid is the most abundant *unsaturated fatty acid* in milk with one double bond. While the cis form of geometric isomer is the most common found in nature, approximately 5% of all unsaturated bonds are in the trans position as a result of rumen hydrogenation.

Lipid Structures

Triglycerides account for 98.3% of milk fat. The distribution of fatty acids on the triglyceride chain, while there are hundreds of different combinations, is not random. The fatty acid pattern is important when determining the physical properties of the lipids. In general, the SN1 position binds mostly longer carbon length fatty acids, and the SN3 position binds mostly shorter carbon length and unsaturated fatty acids. For example:

- C4 - 97% in SN3
- C6 - 84% in SN3
- C18 - 58% in SN1

The small amounts of mono- , diglycerides, and free fatty acids in fresh milk may be a product of early lipolysis or simply incomplete synthesis. Other classes of lipids include phospholipids (0.8%) which are mainly associated with the fat globule membrane, and cholesterol (0.3%) which is mostly located in the fat globule core.

Milk Lipids - Physical Properties

The physical properties of milkfat can be summerized as follows:

- density at 20° C is 915 kg m^{-3}
- refractive index (589 nm) is 1.462 which decreases with increasing temperature
- solubility of water in fat is 0.14% (w/w) at 20°C and increases with increasing temperature
- thermal conductivity is about 0.17 J m^{-1} s^{-1} K^{-1} at 20°C
- specific heat at 40° C is about 2.1kJ kg^{-1} K^{-1}
- electrical conductivity is <10(-12) ohm^{-1} cm^{-1}
- dielectric constant is about 3.1.

At room temperature, the lipids are solid, therefore, are correctly referred to as "fat" as opposed to "oil" which is liquid at room temperature. The *melting points* of individual triglycerides ranges from -75° C for tributyric glycerol to 72° C for tristearin. However, the final melting point of milkfat is at 37° C because higher melting triglycerides dissolve in the liquid fat. This temperature is significant because 37° C is the body temperature of the cow and the milk would need to be liquid at this temperature. The melting curves of milkfat are complicated by the diverse lipid composition:

- trans unsaturation increases melting points;
- odd-numbered and branched chains decrease melting points.

Crystallization

Crystallization of milkfat largely determines the physical stability of the fat globule and the consistency of high-fat dairy products, but crystal behaviour is also complicated by the wide range of different triglycerides. There are four forms that milkfat crystals can occur in; alpha, α , $ß'^{1}$, and $ß'^{2}$, however, the alpha form is the least stable and is rarely observed in slowly cooled fat.

MILKFAT STRUCTURE—FAT GLOBULES

More than 95% of the total milk lipid is in the form of a globule ranging in size from 0.1 to 15 um in diameter. These liquid fat droplets are covered by a thin membrane, 8 to 10 µm in thickness, whose properties are completely different from both milkfat and plasma. The native fat globule membrane (FGM) is comprised of apical plasma membrane of the secretory cell which continually envelopes the lipid droplets as they pass into the lumen. The major components of the native FGM, therefore, is protein and phospholipids. The phospholipids are involved in the oxidation of milk. There may be some rearrangement of the membrane after release into the lumen as amphiphilic substances from the plasma adsorb onto the fat globule and parts of the membrane dissolve into either the globule core or the serum. The FGM decreases the lipid-serum interface to very low values, 1 to 2.5 mN/m, preventing the globules from immediate flocculation and coalescence, as well as protecting them from enzymatic action.

It is well known that if raw milk or cream is left to stand, it will separate. Stokes' Law predicts that fat globules will cream due to the differences in densities between the fat and plasma phases of milk. However, in cold raw milk, creaming takes place faster than is predicted from this fact alone. IgM, an immunoglobulin in milk, forms a complex with lipoproteins. This complex, known as cryoglobulin precipitates onto the fat globules and causes flocculation. This is known as cold agglutination. As fat globules cluster, the speed of rising increases and sweeps up the smaller globules with them. The cream layer forms very rapidly, within 20 to 30 min., in cold milk.

Homogenization of milk prevents this creaming by decreasing the diameter and size distribution of the fat globules, causing the speed of rise to be similar for the majority of globules. As well, homogenization causes the formation of a recombined membrane which is much similar in density to the continuous phase.

RECOMBINED MEMBRANES

Recombined membranes are very different than native FGM. Processing steps such as homogenization, decreases the average diameter of fat globule and significantly increases the surface area. Some of the native FGM will remain adsorbed but there is no longer enough of it to cover all of the newly created surface area. Immediately after disruption of the fat globule, the surface tension raises to a high level of 15 mN/m and amphiphilic molecules in the plasma quickly adsorb to the lipid droplet to lower this value. The adsorbed layers consist mainly of serum proteins and casein micelles.

Fat Destabilization

While homogenization is the principal method for acheiving stabilization of the fat emulsion in milk, fat destabilization is necessary for structure formation inbutter, whipping cream and ice cream. Fat destabilization refers to the process of clustering and clumping (partial coalescence) of the fat globules which leads to the development of a continuous internal fat network or matrix structure in the product. Fat destabilization (sometimes "fat agglomeration") is a general term that describes the summation of several different phenomena. These include:

- *Coalescence:* an irreversible increase in the size of fat globules and a loss of identity of the coalescing globules;
- *Flocculation:* a reversible (with minor energy input) agglomeration/clustering of fat globules with no loss of identity of the globules in the floc; the fat globules that flocculate ; they can be easily redispersed if they are held together by weak forces, or they might be harder to redisperse to they share part of their interfacial layers;
- *Partial coalescence:* an irreversible agglomeration/ clustering of fat globules, held together by a combination of fat crystals and liquid fat, and a retention of identity of individual globules as long as the crystal structure is maintained (i.e., temperature dependent, once the crystals melt, the cluster coalesces). They usually come together in a shear field, as in whipping, and it is envisioned that the crystals at the surface of the droplets

are responsible for causing colliding globules to stick together, while the liquid fat partially flows between they and acts as the "cement". Partial coalescence dominates structure formation in whipped, aerated dairy emulsions, and it should be emphasized that crystals within the emulsion droplets are responsible for its occurrence.

Milk Lipids Functional Properties

Like all fats, milkfat provides lubrication. They impart a creamy mouth feel as opposed to a dry texture. Butter flavour is unique and is derived from low levels of short chain fatty acids. If too many short chain fatty acids are hydrolyzed (separated) from the triglycerides, however, the product will taste rancid. Butter fat also acts as a reservoir for other flavours, especially in aged cheese. Fat globules produce a 'shortening' effect in cheese by keeping the protein matrix extended to give a soft texture. Fat substitutes are designed to mimic the globular property of milk fat. The spreadable range of butter fat is 16-24° C. Unfortunately butter is not spreadable at refrigeration temperatures. Milk fat provides energy (1 g = 9 cal.), and nutrients (essential fatty acids, fat soluble vitamins).

Milk Proteins

The primary structure of proteins consists of a polypeptide chain of amino acids residues joined together by peptide linkages, which may also be cross-linked by disulphide bridges. Amino acids contain both a weakly basic amino group, and a weakly acid carboxyl group both connected to a hydrocarbon chain, which is unique to different amino acids. The three-dimensional organization of proteins, or conformation, also involves secondary, tertiary, and quaternary structures. The secondary structure refers to the spatial arrangement of amino acid residues that are near one another in the linear sequence. The alpha-helix and ß-pleated sheat are examples of secondary structures arising from regular and periodic steric relationships. The tertiary structure refers to the spatial arrangement of amino acid residues that are far apart in the linear sequence, giving rise to further coiling and folding. If the protein is tightly coiled and folded into a somewhat spherical shape, it is called a globular protein. If the protein consists of long polypeptide chains which are intermolecularly linked, they are called fibrous proteins. Quaternary structure occurs when proteins with two or more polypeptide chain subunits are associated.

Milk Protein Fractionation

The nitrogen content of milk is distributed among caseins (76%), whey proteins (18%), and non-protein nitrogen (NPN) (6%). This does not include the minor proteins that are associated with the FGM. This nitrogen distribution can be determined by the Rowland fractionation method:

Precipitation at pH 4.6—Separates Caseins from whey Nitrogen

Precipitation with sodium acetate and acetic acid (pH 5.0) - separates total proteins from whey NPN

The concentration of proteins in milk is as follows:

		grams/ litre	% of total protein
Total Protein		33	100
Total Caseins		26	79.5
	alpha s1	10	30.6
	alpha s2	2.6	8.0
	beta	9.3	28.4
	kappa	3.3	10.1
Total Whey Proteins		6.3	19.3
	alpha lactalbumin	1.2	3.7
	beta lactoglobulin	3.2	9.8
	BSA	0.4	1.2
	Immunoglobulins	0.7	2.1
	Proteose peptone	0.8	2.4

Caseins, as well as their structural form - casein micelles, whey proteins, and milk enzymes will now be examined in further detail.

Caseins

The casein content of milk represents about 80% of milk proteins. The principal casein fractions are alpha (s1) and alpha(s2)-caseins, ß-casein, and kappa-casein. The distinguishing property of all caseins is their low solubility at pH 4.6. The common compositional factor is that caseins are conjugated proteins, most with phosphate group(s) esterified to serine residues. These phosphate groups are important to the structure of the casein micelle. Calcium binding by the individual caseins is proportional to the phosphate content.

The conformation of caseins is much like that of denatured globular proteins. The high number of proline residues in caseins causes particular bending of the protein chain and inhibits the formation of close-packed, ordered secondary structures. Caseins contain no disulfide bonds. As well, the lack of tertiary structure accounts for the stability of caseins against heat denaturation because there is very little structure to unfold. Without a tertiary structure there is considerable exposure of hydrophobic residues. This results in strong association reactions of the caseins and renders them insoluble in water. Within the group of caseins, there are several distinguishing features based on their charge distribution and sensitivity to calcium precipitation:

- alpha(s1)-casein: (molecular weight 23,000; 199 residues, 17 proline residues)

  Two hydrophobic regions, containing all the proline residues, separated by a polar region, which contains all but one of eight phosphate groups. It can be precipitated at very low levels of calcium.

- alpha(s2)-casein: (molecular weight 25,000; 207 residues, 10 prolines)

  Concentrated negative charges near N-terminus and positive charges near C-terminus. It can also be precipitated at very low levels of calcium.

- ß-casein: (molecular weight 24,000; 209 residues, 35 prolines)

Highly charged N-terminal region and a hydrophobic C-terminal region. Very amphiphilic protein acts like a detergent molecule. Self association is temperature dependant; will form a large polymer at 20º C but not at 4º C. Less sensitive to calcium precipitation.

- kappa-casein: (molecular weight 19,000; 169 residues, 20 prolines)

  Very resistant to calcium precipitation, stabilizing other caseins. Rennet cleavage at the Phe105-Met106 bond eliminates the stabilizing ability, leaving a hydrophobic portion, para-kappa-casein, and a hydrophilic portion called kappa-casein glycomacropeptide (GMP), or more accurately, caseinomacropeptide (CMP).

Most, but not all, of the casein proteins exist in a colloidal particle known as the casein micelle. Its biological function is to carry large amounts of highly insoluble CaP to mammalian young in liquid form and to form a clot in the stomach for more efficient nutrition. Besides casein protein, calcium and phosphate, the micelle also contains citrate, minor ions, lipase and plasmin enzymes, and entrapped milk serum. These micelles are rather porous structures, occupying about 4 ml/g and 6-12% of the total volume fraction of milk.

Casein "Sub-micelle"

The "casein sub-micelle" model has been prominent for the last several years, and is illustrated and described with the following link, but there is not universal acceptance of this model, and mounting research evidence to suggest that there is not a defined sub-micellar structure to the micelle at all. Another model of a more open structure is also defined with the following link.

In the submicelle model, it is thought that there are small aggregates of whole casein, containing 10 to 100 casein molecules, called submicelles. It is thought that there are two different kinds of submicelle; with and without kappa-casein. These submicelles contain a hydrophobic core and are covered by a hydrophilic coat which is at least partly comprised of the polar moieties of kappa-casein. The hydrophilic CMP of the kappa-casein exists as a flexible hair.

The open model also suggests there are more dense and less dense regions within the midelle, but there is less of a well-defined structure. In this model, calcium phosphate nanoclusters bind caseins and provide for the differences in density within the casein micelle.

Colloidal Calcium Phosphate

Colloidal calcium phosphate (CCP) acts as a cement between the hundreds or even thousands of submicelles that form the casein micelle. Binding may be covalent or electrostatic. Submicelles rich in kappa-casein occupy a surface position, whereas those with less are buried in the interior. The resulting hairy layer, at least 7 nm thick, acts to prohibit further aggregation of submicelles by steric repulsion. The casein micelles are not static; there are three dynamic equilibria between the micelle and its surroundings:

- the free casein molecules and submicelles
- the free submicelles and micelles
- the dissoved colloidal calcium and phosphate

The following factors must be considered when assessing the stability of the casein micelle:

Role of Ca^{++}: More than 90% of the calcium content of skim milk is associated in some way or another with the casein micelle. The removal of Ca^{++} leads to reversible dissociation of ß-casein without micellular disintegration. The addition of Ca^{++} leads to aggregation.

H Bonding: Some occurs between the individual caseins in the micelle but not much because there is no secondary structure in casein proteins.

Disulphide Bonds: alpha (s1) and ß-caseins do not have any cysteine residues. If any S-S bonds occur within the micelle, they are not the driving force for stabilization.

Hydrophobic Interactions

Caseins are among the most hydrophobic proteins and there is some evidence to suggest they play a role in the stability of the micelle. It must be remembered that hydrophobic interactions are very temperature sensitive.

Electrostatic Interactions

Some of the subunit interactions may be the result of ionic bonding, but the overall micellar structure is very loose and open.

- van der Waals Forces
- No sucess in relating these forces to micellular stability.

Steric Stabilization

As already noted, the hairy layer interferes with interparticle approach.

There are several factors that will affect the stability of the casein micelle system

Salt Content

affects the calcium activity in the serum and calcium phosphate content of the micelles.

pH

Lowering the pH leads to dissolution of calcium phosphate until, at the isoelectric point (pH 4.6), all phosphate is dissolved and the caseins precipitate.

Temperature

At 4° C, beta-casein begins to dissociate from the micelle, at 0° C, there is no micellar aggregation; freezing produces a precipitate called cryo-casein.

Heat Treatment

Whey proteins become adsorbed, altering the behaviour of the micelle.

Dehydration

By ethanol, for example, leads to aggregation of the micelles. When two or more of these factors are applied together, the effect can also be additive.

Casein Micelle Aggregation

Caseins are able to aggregate if the surface of the micelle is reactive. Casein micelle aggregation 17 KB. Although the casein micelle is fairly stable, there are four major ways in which aggregation can be induced:

- chymosin - rennet or other proteolytic enzymes as in Cheese manufacturing
- acid
- heat
- age gelation

Enzyme Coagulation

Chymosin, or rennet, is most often used for enzyme coagulation. During the primary stage, rennet cleaves the Phe(105)-Met(106) linkage of kappa-casein resulting in the formation of the soluble CMP which diffuses away from the micelle and para-kappa-casein, a distinctly hydrophobic peptide that remains on the micelle. The patch or reactive site, as illustrated in the above image, that is left on the micelles after enzymatic cleavage is necessary before aggregation of the paracasein micelles can begin.

During the secondary stage, the micelles aggregate. This is due to the loss of steric repulsion of the kappa-casein as well as the loss of electrostatic repulsion due to the decrease in pH. As the pH approaches its isoelectric point (pH 4.6), the caseins aggregate. The casein micelles also have a strong tendency to aggregate because of hydrophobic interactions. Calcium assists coagulation by creating isoelctric conditions and by acting as a bridge between micelles. The temperature at the time of coagulation is very important to both the primary and secondary stages. With an increase in temperature up to 40° C, the rate of the rennet reaction increases. During the secondary stage, increased temperatures increase the hydrophobic reaction. The tertiary stage of coagulation involves the rearrangement of micelles after a gel has formed. There is a loss of paracasein identity as te milk curd firms and syneresis begins.

Acid Coagulation

Acidification causes the casein micelles to destabilize or aggregate by decreasing their electric charge to that of the isoelectric point. At the same time, the acidity of the medium increases the solubility of minerals so that organic calcium and phosphorus contained in the micelle gradually become soluble in the aqueous phase. Casein micelles disintegrate and casein precipitates. Aggregation occurs as a result of entropically driven hydrophobic interactions.

Heat

At temperatures above the boiling point casein micelles will irreversibly aggregate. On heating, the buffer capacity of milk salts change, carbon dioxide is released, organic acids are produced, and tricalcium phophate and casein phosphate may be precipitated with the release of hydrogen ions.

Age Gelation

Age gelation is an aggregation phenomenon that affects shelf-stable, sterilized dairy products, such as concentrated milk and UHT milk products. After weeks to months storage of these products, there is a sudden sharp increase in viscosity accompanied by visible gelation and irreversible aggregation of the micelles into long chains forming a three-dimensional network. The actual cause and mechanism is not yet clear, however, some theories exist:

Proteolytic breakdown of the casein: bacterial or native plasmin enzymes that are resistant to heat treatment may lead to the formation of a gel.

Chemical reactions: Polymerization of casein and whey proteins due to Maillard type or other chemical reactions.

Formation of kappa-casein-ß -lactoglobulin complexes

An excellent source of information on casein micelle stability can be found in Walstra.

Whey Proteins

The proteins appearing in the supernatant of milk after precipitation at pH 4.6 are collectively called whey proteins.

These globular proteins are more water soluble than caseins and are subject to heat denaturation. Native whey proteins have good gelling and whipping properties. Denaturation increases their water holding capacity. The principle fractions are ß -lactoglobulin, alpha-lactalbumin, bovine serum albumin (BSA), and immunoglobulins (Ig).

ß -Lactoglobulins: (MW - 18,000; 162 residues)

This group, including eight genetic variants, comprises approximately half the total whey proteins. ß-Lactoglobulin has two internal disulfide bonds and one free thiol group. The conformation includes considerable secondary structure and exists naturally as a noncovalent linked dimer. At the isoelectric point (pH 3.5 to 5.2), the dimers are further associated to octamers but at pH below 3.4, they are dissociated to monomers.

Alpha-Lactalbumins: (MW - 14,000; 123 residues)

These proteins contain eight cysteine groups, all involved in internal disulfide bonds, and four tryptophan residues. Alpha-Lactalbumin has a highly ordered secondary structure, and a compact, spherical tertiary structure. Thermal denaturation and pH <4.0 results in the release of bound calcium.

Enzymes

Enzymes are a group of proteins that have the ability to catalyze chemical reactions and the speed of such reactions. The action of enzymes is very specific. Milk contains both indigenous and exogenous enzymes. Exogenous enzymes mainly consist of heat-stable enzymes produced by psychrotrophic bacteria: lipases, and proteinases. There are many indigenous enzymes that have been isolated from milk. The most significant group are the hydrolases:

- lipoprotein lipase
- plasmin
- alkaline phosphatase

Lipoprotein lipase (LPL)

A lipase enzyme splits fats into glycerol and free fatty acids. This enzyme is found mainly in the plasma in association with casein micelles. The milkfat is protected from its action by the FGM. If the FGM has been damaged, or if certain cofactors (blood serum lipoproteins) are present, the LPL is able to attack the lipoproteins of the FGM. Lipolysis may be caused in this way.

Plasmin

Plasmin is a proteolytic enzyme; it splits proteins. Plasmin attacks both ß-casein and alpha(s2)-casein. It is very heat stable and responsible for the development of bitterness in pasteurized milk and UHT processed milk. It may also play a role in the ripening and flavour development of certain cheeses, such as Swiss cheese.

Alkaline Phosphatase

Phosphatase enzymes are able to split specific phosporic acid esters into phosphoric acid and the related alcohols. Unlike most milk enzymes, it has a pH and temperature optima differing from physiological values; pH of 9.8. The enzyme is destroyed by minimum pasteurization temperatures, therefore, a phosphatase test can be done to ensure proper pasteurization. One of its most important functions is its utilization as a fermentation substrate. Lactic acid bacteria produce lactic acid from lactose, which is the beginning of many fermented dairy products. Because of their ability to metabolize lactose, they have a competitive advantage over many pathogenic and spoilage organisms. Some people suffer from lactose intolerance; they lack the lactase enzyme, hence they cannot digest lactose, or dairy products containing lactose. Crystallization of lactose occurs in an alpha form which commonly takes a tomahawk shape. This results in the defect called sandiness. Lactose is relatively insoluble which is a problem in many dairy products, ice cream, sweetened condensed milk. In addition to lactose, fresh milk contains other carbohydrates in small amounts, including glucose, galactose, and oligosaccharides.

Vitamins

Vitamins are organic substances essential for many life processes. Milk includes fat soluble vitamins A , D, E, and K. Vitamin A is derived from retinol and ß -carotene. Because milk is an important source of dietary vitamin A, fat reduced products which have lost vitamin A with the fat, are required to supplement the product with vitamin A. Milk is also an important source of dietary water soluble vitamins:

- B_1 - thiamine
- B_2 - riboflavin
- B_6 - pyridoxine
- B_{12} - cyanocobalamin
- niacin
- pantothenic acid.

There is also a small amount of vitamin C (ascorbic acid) present in raw milk but is very heat-labile and easily destroyed by pasteurization. The vitamin content of fresh milk is given below:

Vitamin	Contents per litre
A (ug RE)	400
D (IU)	40
E (µg)	1000
K (µg)	50
B1 (µg)	450
B2 (µg)	1750
Niacin (µg)	900
B6 (µg)	500
Pantothenic acid (µg)	3500
Biotin (µg)	35
Folic acid (µg)	55
B12 (µg)	4.5
C (mg)	20

Minerals

All 22 minerals considered to be essential to the human diet are present in milk. These include three families of salts:

- *Sodium (Na), Potassium (K) and Chloride (Cl):* These free ions are negatively correlated to lactose to maintain osmotic equilibrium of milk with blood.
- *Calcium (Ca), Magnesium (Mg), Inorganic Phosphorous (P(i)), and Citrate*: This group consists of 2/3 of the Ca, 1/3 of the Mg, 1/2 of the P(i), and less than 1/10 of the citrate in colloidal (nondiffusible) form and present in the casein micelle.
- *Diffusible salts of Ca, Mg, citrate, and phosphate:* These salts are very pH dependent and contribute to the overall acid-base equilibrium of milk.

The mineral content of fresh milk is given below:

Mineral	Content per litre
Sodium (mg)	350-900
Potassium (mg)	1100-1700
Chloride (mg)	900-1100
Calcium (mg)	1100-1300
Magnesium (mg)	90-140
Phosphorus (mg)	900-1000
Iron (µg)	300-600
Zinc (µg)	2000-6000
Copper (µg)	100-600
Manganese (µg)	20-50
Iodine (µg)	260
Fluoride (µg)	30-220
Selenium (µg)	5-67
Cobalt (µg)	0.5-1.3
Chromium (µg)	8-13
Molybdenum (µg)	18-120
Nickel (µg)	0-50

Silicon (µg)	750-7000
Vanadium (µg)	tr-310
Tin (µg)	40-500
Arsenic (µg)	20-60

Physical Properties

Density

The density of milk and milk products is used for the following:

- to convert volume into mass and *vice versa;*
- to estimate the solids content; and
- to calculate other physical properties (e.g. kinematic viscosity).

Density, the mass of a certain quantity of material divided by its volume, is dependant on the following:

- temperature at the time of measurement;
- temperature history of the material;
- composition of the material (especially the fat content);
- inclusion of air (a complication with more viscous products).

With all of this in mind, the density of milk varies within the range of 1027 to 1033 kg m(-3) at 20° C.

The following table gives the density of various fluid dairy products as a function of fat and solids-not-fat (SNF) composition:

Viscosity

Viscosity of milk and milk products is important in determining the following:

- the rate of creaming'
- rates of mass and heat transfer; and
- the flow conditions in dairy processes.

Density of Various Fluid Dairy Products

Product	Project Composition		Density (Kg/L) at			
	Fat%	SNF(%)	4.4°C	10°C	20°C	38.9°C
Producer milk	4.00	8.95	1.035	1.033	1.030	1.023
Homogenized milk	3.6	8.6	1.033	1.032	1.029	1.022
Skim milk, pkg	0.02	8.9	1.036	1.035	1.033	1.026
Fortified skim	0.02	10.15	1.041	1.040	1.038	1.031
Half and half	12.25	7.75	1.027	1.025	1.020	1.010
Half and half, fort.	11.30	8.9	1.031	1.030	1.024	1.014
Light cream	20.00	7.2	1.021	1.018	1.012	1.000
Heavy cream	36.60	5.55	1.008	1.005	0.994	0.978

Milk and skim milk, excepting cooled raw milk, exhibit Newtonian behavior, in which the viscosity is independent of the rate of shear. The viscosity of these products depends on the following:

- *Temperature:* cooler temperatures increase viscosity due to the increased voluminosity of casein micelles temperatures above 65° C increase viscosity due to the denaturation of whey proteins
- *pH:* an increase or decrease in pH of milk also causes an increase in casein micelle voluminosity.

Cooled raw milk and cream exhibit non-Newtonian behaviour in which the viscosity is dependant on the shear rate. Agitation may cause partial coalescence of the fat globules (partial churning) which increases viscocity. Fat globules that have under gone cold agglutination, may be dispersed due to agitation, causing a decrease in viscosity.

Freezing Point

Freezing point depression is a colligative property which is determined by the molarity of solutes rather than by the percentage by weight or volume. In the dairy industry, freezing point of milk is mainly used to determine added water but it can also been used to determine lactose content in milk, estimate whey powder contents in skim milk powder, and to determine water activity of cheese. The freezing point of milk is usually in the range of -0.512 to -0.550° C with an average of about -0.522° C.

Correct interpretation of freezing point data with respect to added water depends on a good understanding of the factors affecting freezing point depression. With respect to interpretation of freezing points for added water determination, the most significant variables are the nutritional status of the herd and the access to water. Under feeding causes increased freezing points. Large temporary increases in freezing point occur after consumption of large amounts of water because milk is iso-osmotic with blood. The primary sources of non-intentional added water in milk are residual rinse water and condensation in the milking system.

Acid-Base Equilibria

Both titratable acidity and pH are used to measure milk acidity. The pH of milk at 25° C normally varies within a relatively narrow range of 6.5 to 6.7. The normal range for titratable acidity of herd milks is 13 to 20 mmol/L. Because of the large inherent variation, the measure of titratable acidity has little practical value except to measure changes in acidity (e.g., during lactic fermentation) and even for this purpose, pH is a better measurement.

There are many components in milk which provide a buffering action. The major buffering groups of milk are caseins and phosphate.

Optical Properties

Optical properties provide the basis for many rapid, indirect methods of analysis such as proximate analysis by infrared absorbency or light scattering. Optical properties also determine the appearance of milk and milk products. Light scattering by fat globules and casein micelles causes milk to appear turbid and opaque. Light scattering occurs when the wave length of light is near the same magnitude as the particle. Thus, smaller particles scatter light of shorter wavelengths. Skim milk appears slightly blue because casein micelles scatter the shorter wavelengths of visible light (blue) more than the red. The carotenoid precursor of vitamin A, ß-carotene, contained in milk fat, is responsible for the 'creamy' colour of milk. Riboflavin imparts a greenish colour to whey.

Refractive index (RI) is normally determined at 20°C with the D line of the sodium spectrum. The refractive index of milk is 1.3440 to 1.3485 and can be used to estimate total solids.

5

DAIRY MICROBIOLOGY

Intoduction

Microorganisms are living organisms that are individually too small to see with the naked eye. The unit of measurement used for microorganisms is the micrometre (µm); 1 µm = 0.001 millimetre; 1 nanometre (nm) = 0.001 µm. Microorganisms are found everywhere (ubiquitous) and are essential to many of our planets life processes. With regards to the food industry, they can cause spoilage, prevent spoilage through fermentation, or can be the cause of human illness.

There are several classes of microorganisms, of which bacteria and fungi (yeasts and moulds) will be discussed in some detail. Another type of microorganism, the bacterial viruses or bacteriophage, will be examined in a later section.

Bacteria

Bacteria are relatively simple single-celled organisms. One method of classification is by shape or morphology.

Cocci

— spherical shape

— 0.4 - 1.5 µm

Examples: staphylococci - form grape-like clusters; streptococci - form bead-like chains.

Rods

— 0.25 - 1.0 µm width by 0.5 - 6.0 µm long

Examples: bacilli - straight rod; *spirilla* - spiral rod

There exists a bacterial system of taxonomy, or classification system, that is internationally recognized with family, genera and species divisions based on genetics.

Some bacteria have the ability to form resting cells known as endospores. The spore forms in times of environmental stress, such as lack of nutrients and moisture needed for growth, and thus is a survival strategy. Spores have no metabolism and can withstand adverse conditions such as heat, disinfectants, and ultraviolet light. When the environment becomes favourable, the spore germinates and giving rise to a single vegetative bacterial cell. Some examples of spore-formers important to the food industry are members of *Bacillus* and *Clostridium* generas.

Bacteria reproduce asexually by fission or simple division of the cell and its contents. The doubling time, or generation time, can be as short as 20-20 min. Since each cell grows and divides at the same rate as the parent cell, this could under favourable conditions translate to an increase from one to 10 million cells in 11 hours! However, bacterial growth in reality is limited by lack of nutrients, accumulation of toxins and metabolic wastes, unfavourable temperatures and dessication. The maximum number of bacteria is approximately 1 × 10e9 CFU/g or ml.

Note: Bacterial populations are expressed as colony forming units (CFU) per gram or millilitre.

Bacterial Growth

Bacterial growth generally proceeds through a series of phases:

- *Lag phase:* time for microorganisms to become accustomed to their new environment. There is little or no growth during this phase.

- *Log phase:* bacteria logarithmic, or exponential, growth begins; the rate of multiplication is the most rapid and constant.
- *Stationary phase:* the rate of multiplication slows down due to lack of nutrients and build-up of toxins. At the same time, bacteria are constantly dying so the numbers actually remain constant.
- *Death phase:* cell numbers decrease as growth stops and existing cells die off.

The shape of the curve varies with temperature, nutrient supply, and other growth factors. This exponential death curve is also used in modelling the heating destruction of microorganisms.

Yeasts

Yeasts are members of a higher group of microorganisms called fungi. They are single-cell organisms of spherical, elliptical or cylindrical shape. Their size varies greatly but are generally larger than bacterial cells. Yeasts may be divided into two groups according to their method of reproduction:

1. *budding:* called Fungi Imperfecti or false yeasts;
2. budding and spore formation: called Ascomycetes or true yeasts.

Unlike bacterial spores, yeast form spores as a method of reproduction.

Moulds

Moulds are filamentous, multi-celled fungi with an average size larger than both bacteria and yeasts (10 X 40 µ m). Each filament is referred to as a *hypha*. The mass of hyphae that can quickly spread over a food substrate is called the *mycelium*. Moulds may reproduce either asexually or sexually, sometimes both within the same species.

Asexual Reproduction

- *fragmentation* - hyphae separate into individual cells called arthropsores.

- *spore production* - formed in the tip of a fruiting hyphae, called conidia, or in swollen structures called sporangium.

  Sexual Reproduction: sexual spores are produced by nuclear fission in times of unfavourable conditions to ensure survival.

Microbial Growth

There are a number of factors that affect the survival and growth of microorganisms in food. The parameters that are inherent to the food, or *intrinsic factors*, include the following:

- nutrient content
- moisture content
- pH
- available oxygen
- biological structures
- antimicrobial constituents

Nutrient Requirements

While the nutrient requirements are quite organism specific, the microorganisms of importance in foods require the following:

- water
- energy source
- carbon/nitrogen source
- vitamins
- minerals

Milk and dairy products are generally very rich in nutrients which provides an ideal growth environment for many microorganisms.

Moisture Content

All microorganisms require water but the amount necessary for growth varies between species. The amount of water that is available in food is expressed in terms of water activity (aw), where the aw of pure water is 1.0. Each microorganism has a maximum, optimum, and minimum aw for growth and survival. Generally bacteria dominate in foods with high aw (minimum approximately 0.90 aw) while yeasts and moulds, which require less moisture, dominate in low aw foods (minimum 0.70 aw). The water activity of fluid milk is approximately 0.98 aw.

pH

Most microorganisms have approximately a neutral pH optimum (pH 6-7.5). Yeasts are able to grow in a more acid environment compared to bacteria. Moulds can grow over a wide pH range but prefer only slightly acid conditions. Milk has a pH of 6.6 which is ideal for the growth of many microoorganisms.

Available Oxygen

Microorganisms can be classified according to their oxygen requirements necessary for growth and survival:

- *Obligate Aerobes:* oxygen required
- *Facultative:* grow in the presence or absence of oxygen.
- *Microaerophilic:* grow best at very low levels of oxygen.
- *Aerotolerant Anaerobes:* oxygen not required for growth but not harmful if present.
- *Obligate Anaerobes:* grow only in complete absence of oxygen; if present it can be lethal.

Biological Structures

Physical barriers such as skin, rinds, feathers, etc. have provided protection to plants and animals against the invasion of microorganisms. Milk, however, is a fluid product with no barriers to the spreading of microorganisms throughout the product.

Antimicrobial Constituents

As part of the natural protection against microorganisms,

many foods have antimicrobial factors. Milk has several nonimmunological proteins which inhibit the growth and metabolism of many microorganisms including the following most common:

- lactoperoxidase
- lactoferrin
- lysozyme
- xanthine

More information on these antimicrobials can be found in a chapter on dairy microbiology and safety written by *Vasavada* and *Cousin.*

Where the intrinsic factors are related to the food properties, the extrinsic factors are related to the storage environment. These would include temperature, relative humidity, and gases that surround the food.

Temperature

As a group, microorganisms are capable of growth over an extremely wide temperature range. However, in any particular environment, the types and numbers of microorganisms will depend greatly on the temperature. According to temperature, microorganisms can be placed into one of three broad groups:

- *Psychrotrophs:* optimum growth temperatures 20 to 30° capable of growth at temperatures less than 7° C. Psychrotrophic organisms are specifically important in the spoilage of refrigerated dairy products.
- *Mesophiles:* optimum growth temperatures 30 to 40° C; do not grow at refrigeration temperatures.
- *Thermophiles:* optimum growth between 55 and 65° C

It is important to note that for each group, the growth rate increases as the temperature increases only up to an optimum, afterwhich it rapidly declines.

Detection and Enumeration of Microorganisms

There are several methods for detection and enumeration of microorganisms in food. The method that is used depends on the purpose of the testing.

Direct Enumeration

Using direct microscopic counts (DMC), Coulter counter etc. allows a rapid estimation of all viable and nonviable cells. Identification through staining and observation of morphology also possible with DMC.

Viable Enumeration

The use of standard plate counts, most probable number (MPN), membrane filtration, plate loop methos, spiral plating etc., allows the estimation of only viable cells. As with direct enumeration, these methods can be used in the food industry to enumerate fermentation, spoilage, pathogenic, and indicator organisms.

Metabolic Activity Measurement

An estimation of metabolic activity of the total cell population is possible using dye reduction tests such as resazurin or methylene blue dye reduction, acid production, electrical impedence etc. The level of bacterial activity can be used to assess the keeping quality and freshness of milk. Toxin levels can also be measured, indicating the presence of toxin producing pathogens.

Cellular Constituents Measurement

Using the luciferase test to measure ATP is one example of the rapid and sensitive tests available that will indicate the presence of even one pathogenic bacterial cell. Isolation of microorganisms is an important preliminary step in the identification of most food spoilage and pathogenic organisms. This can be done using a simple streak plate method.

MICROORGANISMS IN MILK

Milk is sterile at secretion in the udder but is contaminated by bacteria even before it leaves the udder.

Except in the case of mastisis, the bacteria at this point are harmless and few in number. Further infection of the milk by microorganisms can take place during milking, handling, storage, and other pre-processing activities.

Lactic Acid Bacteria

This group of bacteria are able to ferment lactose to lactic acid. They are normally present in the milk and are also used as starter cultures in the production of cultured dairy products such as yogurt.

Note: many lactic acid bacteria have recently been reclassified; the older names will appear in brackets as you will still find the older names used for convenience sake in a lot of literature. Some examples in milk are:

- lactococci
- *L. delbrueckii subsp.* lactis (*Streptococcus lactis*)
- Lactococcus lactis subsp. *cremoris (Streptococcus cremoris*)
- *Lactobacilli*
- *Lactobacillus* casei
- *L.delbrueckii* subsp. lactis (L. lactis)
- *L. delbrueckii* subsp. *bulgaricus* (*Lactobacillus bulgaricus*)
- *Leuconostoc*

Coliforms

Coliforms are facultative anaerobes with an optimum growth at 37° C. Coliforms are indicator organisms; they are closely associated with the presence of pathogens but not necessarily pathogenic themselves. They also can cause rapid spoilage of milk because they are able to ferment lactose with the production of acid and gas, and are able to degrade milk proteins. They are killed by HTST treatment, therefore, their presence after treatment is indicative of contamination.Escherichia coli is an example belonging to this group.

SIGNIFICANCE OF MICROORGANISMS IN MILK

Information on the microbial content of milk can be used to judge its sanitary quality and the conditions of production. If permitted to multiply, bacteria in milk can cause spoilage of the product. Milk is potentially susceptible to contamination with pathogenic microorganisms. Precautions must be taken to minimize this possibility and to destroy pathogens that may gain entrance.

Certain microorganisms produce chemical changes that are desirable in the production of dairy products such as cheese, yogurt.

Spoilage Microorganisms in Milk

The microbial quality of raw milk is crucial for the production of quality dairy foods. Spoilage is a term used to describe the deterioration of a foods' texture, colour, odour or flavour to the point where it is unappetizing or unsuitable for human consumption. Microbial spoilage of food often involves the degradation of protein, carbohydrates, and fats by the microorganisms or their enzymes.

In milk, the microorganisms that are principally involved in spoilage are psychrotrophic organisms. Most psychrotrophs are destroyed by pasteurization temperatures, however, some like *Pseudomonas fluorescens*, *Pseudomonas fragi* can produce proteolytic and lipolytic extracellular enzymes which are heat stable and capable of causing spoilage.

Some species and strains of *Bacillus*, *Clostridium*, *Cornebacterium*, *Arthrobacter*, *Lactobacillus*, *Microbacterium*, *Micrococcus*, and *Streptococcus* can survive pasteurization and grow at refrigeration temperatures which can cause spoilage problems.

PATHOGENIC MICROORGANISMS IN MILK

Hygienic milk production practices, proper handling and storage of milk, and mandatory pasteurization has decreased the threat of milkborne diseases such as tuberculosis, brucellosis, and typhoid fever. There have been a number of foodborne illnesses resulting from the ingestion of raw milk, or dairy products made with milk that was not properly pasteurized or was poorly handled causing post-processing contamination. The following bacterial pathogens are still of concern today in raw milk and other dairy products:

- *Bacillus Cereus*
- *Listeria monocytogenes*
- *Yersinia enterocolitica*
- *Salmonella spp.*
- *Escherichia coli* O157:H7
- *Campylobacter jejuni*

It should also be noted that moulds, mainly of species of *Aspergillus, Fusarium* and *Penicillium* can grow in milk and dairy products. If the conditions permit, these moulds may produce mycotoxins which can be a health hazard.

Hazard Analysis and Critical Control Points (HACCP)

Raw and end-products may be tested for the presence, level, or absence of microorganisms. Traditionally these practices were used to reduce manufacturing defects in dairy products and ensure compliance with specifications and regulations, however, they have many drawbacks:

1. destructive and time consuming;
2. slow response;
3. small sample size; and
4. delays in the release of the food.

In the 1960's, the Pillsbury Company, the U.S. Army, and NASA introduced a system for assuring pathogen-free foods for the space program. This system, called Hazard Analysis and Critical Control Points (HACCP), is a focus on critical food safety areas as part of total quality programs. It involves a critical examination of the entire food manufacturing process to determine every step where there is a possibility of physical, chemical, or microbiological contamination of the food which would render it unsafe or unacceptable for human consumption. These identified points are the critical control points (CCP). There are seven prinicples to HAÇCP:

- analyze hazards;
- determine CCPs;
- establish critical limits;
- establish monitoring procedures;
- establish deviation procedures;
- establish verification procedures;
- establish record keeping procedures

Before these principles can be put into place, a prerequisite programme and preliminary setup is necessary.

Prerequisite Programme

- premise control
- receiving and storage control
- equipment performance and maintenance control
- personnel training
- sanitation
- recall procedure
- Preliminary Setup:
- assemble team
- describe the product
- identify intended use
- construct flow diagram and plant schematic
- verify the diagram on-site

Food Safety Enhancement Programme (FSEP)

The FSEP is Canadian Food Inspection Agency's HACCP initiative. There is extensive information at their Web site regarding FSEP, including implementation manuals, HACCP curriculum guidelines, and generic models.

STARTER CULTURES

Starter cultures are those microorganisms that are used in the production of cultured dairy products such as yogurt and cheese. The natural microflora of the milk is either inefficient, uncontrollable, and unpredictable, or is destroyed altogether by the heat treatments given to the milk. A starter culture can provide particular characteristics in a more controlled and predictable fermentation. The primary function of lactic starters is the production of lactic acid from lactose. Other functions of starter cultures may include the following:

- flavour, aroma, and alcohol production;
- proteolytic and lipolytic activities;
- inhibition of undesirable organisms;

There are two groups of lactic starter cultures:

- *simple or defined:* single strain, or more than one in which the number is known;
- *mixed or compound:* more than one strain each providing its own specific characteristics;

Starter cultures may be categorized as mesophilic or thermophilic:

Mesophilic

- *Lactococcus lactis* subsp. cremoris
- *L. delbrueckii* subsp. *lactis*
- *L. lactis* subsp. *lactis* biovar *diacetylactis*
- *Leuconostoc mesenteroides* subsp. *cremoris*

Thermophilic

- *Streptococcus salivarius* subsp. *thermophilus* (*S.thermophilus*)
- *Lactobacillus delbrueckii* subsp. *bulgaricus*
- *L. delbrueckii subsp. lactis*
- *L. casei*

- *L. helveticus*
- *L. plantarum*

Mixtures of mesophilic and thermophilic microorganisms can also be used as in the production of some cheeses.

Bacteriophage

Bacteriophages are viruses that require bacteria host cells for growth and reproduction. Initially, the bacteriophage attaches itself to the bacteria cell wall and injects nuclear substance into the cell. Inside the cell, the nuclear substance produces shells, or phage coats, for the new bacteriophage which are quickly filled with nucleic acid. The bacterial cell ruptures and dies as the new bacteriophage are released.

Bacteriophages are ubiquitous but generally enter the milk processing plant with the farm milk. They can be inactivated heat treatments of 30 min at 63 to 88° C, or by the use of chemical disinfectants.

Bacteriophages are of most concern in cheese making. They attack and destroy most of the lactic acid bacteria which prevents normal ripening known as slow or dead vat.

Starter Culture Preparation

Commercial manufacturers provide starter cultures in lyophilized (freeze-dryed), frozen or spray-dried forms. The dairy product manufacturers need to inoculate the culture into milk or other suitable substrate. There are a number of steps necessary for the propagation of starter culture ready for production:

1. Commercial Culture
2. *Mother culture* - first inoculation; all cultures will originate from this preparation.
3. *Intermediate culture* - in preparation of larger volumes of prepared starter.
4. *Bulk starter culture* - this stage is used in dairy product production.

6

Dairy Processing

CENTRIFUGATION

Centrifugal separation is a process used quite often in the dairy industry. Some uses include:

- clarification (removal of solid impurities from milk prior to pasteurization)
- skimming (separation of cream from skim milk)
- standardizing
- whey separation (separation of whey cream (fat) from whey)
- bactofuge treatment (separation of bacteria from milk)
- quark separation (separation of quarg curd from whey)
- butter oil purification (separation of serum phase from anhydrous milk fat)

PRINCIPLES OF CENTRIFUGATION

Centrifugation is based on Stoke's Law. The particle sedimentation velocity increases with:

- increasing diameter
- increasing difference in density between the two phases
- decreasing viscosity of the continuous phase.

If raw milk were allowed to stand, the fat globules would begin to rise to the surface in a phenomena called creaming. Raw milk in a rotating container also has centrifugal forces acting on it. This allows rapid separation of milk fat from the skim milk portion and removal of solid impurities from the milk.

Separation

Centrifuges can be used to separate the cream from the skim milk. The centrifuge consists of up to 120 discs stacked together at a 45 to 60 degree angle and separated by a 0.4 to 2.0 mm gap or separation channel. Milk is introduced at the outer edge of the disc stack. The stack of discs has vertically aligned distribution holes into which the milk is introduced.

Under the influence of centrifugal force the fat globules (cream), which are less dense than the skim milk, move inwards through the separation channels toward the axis of rotation. The skim milk will move outwards and leaves through a separate outlet.

Clarification

Separation and clarification can be done at the same time in one centrifuge. Particles, which are more dense than the continuous milk phase, are thrown back to the perimeter. The solids that collect in the centrifuge consist of dirt, epithelial cells, leucocytes, corpuscles, bacteria sediment and sludge. The amount of solids that collect will vary, however, it must be removed from the centrifuge.

More modern centrifuges are self-cleaning allowing a continuous separation/clarification process. This type of centrifuge consists of a specially constructed bowl with peripheral discharge slots. These slots are kept closed under pressure. With a momentary release of pressure, for about 0.15 s, the contents of sediment space are evacuated. This can mean anywhere from 8 to 25 L are ejected at intervals of 60 min. For one dairy, self-cleaning translated to a loss of 50 L/hr of milk. The following image is a schematic of both a clarifier and a separator.

Clarification and Separation: 26 KB

Standardization

The streams of skim and cream after separation must be recombined to a specified fat content. This can be done by adjusting the throttling valve of the cream outlet; if the valve is completely closed, all milk will be discharged through the skim milk outlet. As the valve is progressively opened, larger amounts of cream with diminishing fat contents are discharged from the cream outlet. With direct standardization the cream and skim are automatically remixed at the separator to provide the desired fat content (see diagram to explain this automatic standardization process) . Some basic standardization problems including mass balance and Pearson square approach can be viewed here.

Pasteurization

Introduction

The process of pasteurization was named after Louis Pasteur who discovered that spoilage organisms could be inactivated in wine by applying heat at temperatures below its boiling point. The process was later applied to milk and remains the most important operation in the processing of milk.

Definition

The heating of every particle of milk or milk product to a specific temperature for a specified period of time without allowing recontamination of that milk or milk product during the heat treatment process.

Purpose

There are two distinct purposes for the process of milk pasteurization.

Public Health Aspect

To make milk and milk products safe for human consumption by destroying all bacteria that may be harmful to health (pathogens).

Keeping Quality Aspect

To improve the keeping quality of milk and milk products. Pasteurization can destroy some undesirable enzymes and many spoilage bacteria. Shelf life can be 7, 10, 14 or up to 16 days.

The extent of microorganism inactivation depends on the combination of temperature and holding time. Minimum temperature and time requirements for milk pasteurization are based on thermal death time studies for the most heat resistant pathogen found in milk, Coxelliae burnettii. Thermal lethality determinations require the applications of microbiology to appropriate processing determinations. An overview can be found here.

To ensure destruction of all pathogenic microorganisms, time and temperature combinations of the pasteurization process are highly regulated:

ONTARIO PASTEURIZATION REGULATIONS

Milk

- 63° C for not less than 30 min.
- 72° C for not less than 16 sec.

Or equivalent destruction of pathogens and the enzyme phosphatase as permitted by Ontario Provincial Government authorities. Milk is deemed pasteurized if it tests negative for alkaline phosphatase.

Frozen dairy dessert mix (ice cream or ice milk, egg nog):

- at least 69° C for not less than 30 min;
- at least 80° C for not less than 25 sec;

other time temperature combinations must be approved (e.g. 83° C/16 sec).

Milk Based Products

With 10% mf or higher, or added sugar (cream, chocolate milk, etc).

66° C/30 min, 75° C/16 sec

There has also been some progress with low temperature pasteurization methods using membrane processing technology.

Methods of Pasteurization

There are two basic methods, batch or continuous.

Batch Method

The batch method uses a vat pasteurizer which consists of a jacketed vat surrounded by either circulating water, steam or heating coils of water or steam.

Batch Pasteurizer (26 KB)

In the vat the milk is heated and held throughout the holding period while being agitated. The milk may be cooled in the vat or removed hot after the holding time is completed for every particle. As a modification, the milk may be partially heated in tubular or plate heater before entering the vat. This method has very little use for milk but some use for milk by-products (e.g. creams, chocolate) and special batches. The vat is used extensivly in the ice cream industry for mix quality reasons other than microbial reasons.

Continuous Method

Continuous process method has several advantages over the vat method, the most important being time and energy saving. For most continuous processing, a high temperature short time (HTST) pasteurizer is used. The heat treatment is accomplished using a plate heat exchanger. This piece of equipment consists of a stack of corrugated stainless steel plates clamped together in a frame. There are several flow patterns that can be used. Gaskets are used to define the boundaries of the channels and to prevent leakage. The heating medium can be vacuum steam or hot water.

HTST Milk Flow Overview

This overview is meant as an introduction and a summary. Each piece of HTST equipment will be discussed in further detail later.

Cold raw milk at 4° C in a constant level tank is drawn into the regenerator section of pasteurizer. Here it is warmed to approximately 57° C - 68° C by heat given up by hot pasteurized milk flowing in a counter current direction on the opposite side of thin, stainless steel plates. The raw milk, still under suction, passes through a positive displacement timing pump which delivers it under positive pressure through the rest of the HTST system.

The raw milk is forced through the heater section where hot water on opposite sides of the plates heat milk to a temperature of at least 72° C. The milk, at pasteurization temperature and under pressure, flows through the holding tube where it is held for at least 16 sec. The maximum velocity is governed by the speed of the timing pump, diameter and length of the holding tube, and surface friction. After passing temperature sensors of an indicating thermometer and a recorder-controller at the end of the holding tube, milk passes into the flow diversion device (FDD). The FDD assumes a forward-flow position if the milk passes the recorder-controller at the preset cut-in temperature (>72° C). The FDD remains in normal position which is in diverted-flow if milk has not achieved preset cut-in temperature. The improperly heated milk flows through the diverted flow line of the FDD back to the raw milk constant level tank. Properly heated milk flows through the forward flow part of the FDD to the pasteurized milk regenerator section where it gives up heat to the raw product and in turn is cooled to approximately 32° C - 9° C.

The warm milk passes through the cooling section where it is cooled to 4° C or below by coolant on the opposite sides of the thin, stainless steel plates. The cold, pasteurized milk passes through a vacuum breaker at least 12 inches above the highest raw milk in the HTST system then on to a storage tank filler for packaging.

Basic Flow - HTST Pasteurization 17 KB

Holding Time

When fluids move through a pipe, either of two distinct types of flow can be observed. The first is known as turbulent flow which occurs at high velocity and in which eddies are present moving in all directions and at all angles to the normal line of flow. The second type is streamline, or laminar flow which occurs at low velocities and shows no eddy currents. The Reynolds number , is used to predict whether laminar or turbulent flow will exist in a pipe:

- Re < 2100 laminar
- Re > 4000 fully developed turbulent flow

There is an impact of these flow patterns on holding time calculations and the assessment of proper holding tube lengths. The holding time is determined by timing the interval for an added trace substance (salt) to pass through the holder. The time interval of the fastest particle of milk is desired. Thus the results found with water are converted to the milk flow time by formulation since a pump may not deliver the same amount of milk as it does water.

Pressure Differential

For contiunuous pasteurizing, it is important to maintain a higher pressure on the pasteurized side of the heat exchanger. By keeping the pasteurized milk at least one psi higher than raw milk in regenerator, it prevents contamination of pasteurized milk with raw milk in event that a pin-hole leak develops in thin stainless steel plates. This pressure differential is maintained using a timing pump in simple systems, and differential pressure controllers and back pressure flow regulators at the chilled pasteurization outlet in more complex systems. The position of the timing pump is crucial so that there is suction on the raw regenerator side and pushes milk under pressure through pasteurized regenerator. There are several other factors involved in maintaining the pressure differential:

- The balance tank overflow level must be less than the level of lowest milk passage in the regenerator.
- Properly installed booster pump is all that is permitted between balance tank and raw regenerator.

- No pump after pasteurized milk outlet to vacuum breaker.
- There must be greater than a 12 inch vertical rise to the vacuum breaker.
- The raw regenerator drains freely to balance tank at shut-down.

Basic Component Equipment of HTST Pasteurizer

Balance Tank

The balance, or constant level tank provides a constant supply of milk. It is equipped with a float valve assembly which controls the liquid level nearly constant ensuring uniform head pressure on the product leaving the tank. The overflow level must always be below the level of lowest milk passage in regenerator. It, therefore, helps to maintain a higher pressure on the pasteurized side of the heat exchanger. The balance tank also prevents air from entering the pasteurizer by placing the top of the outlet pipe lower than the lowest point in the tank and creating downward slopes of at least two per cent. The balance tank provides a means for recirculation of diverted or pasteurized milk.

Regenerator

Heating and cooling energy can be saved by using a regenerator which utilizes the heat content of the pasteurized milk to warm the incoming cold milk. Its efficiency may be calculated as follows:

% regeneration = temp. increase due to regenerator/ totaltemp. increase

For example: Cold milk entering system at 4° C, after regeneration at 65° C, and final temperature of 72° C would have an 89.7% regeneration:

$$\frac{65-4}{72-4}=89.7$$

Timing Pump

The timing pump draws product through the raw regenerator and pushes milk under pressure through pasteurized regenerator. It governs the rate of flow through the holding tube. It must be a positive displacement pump equipped with variable speed drive that can be legally sealed at the maximum rate to give minimum holding time in holding tubes. It also must be interwired so it only operates when FDD is fully forward or fully diverted, and must be "fail-safe". A centrifigal pump with magnetic flow meter and controller may also be used.

Holding Tube

Must slope upwards 1/4"/ft. in direction of flow to eliminate air entrapment so nothing flows faster at air pocket restrictions.

Indicating Thermometer

The indicating thermometer is considered the most accurate temperature measurement. It is the official temperature to which the safety thermal limit recorder (STLR) is adjusted. The probe should sit as close as possible to STLR probe and be located not greater than 18 inches upstream of the flow diversion device.

Recorder-controller (STLR)

The STLR records the temperature of the milk and the time of day. It monitors, controls and records the position of the flow diversion device (FDD) and supplies power to the FDD during forward flow. There are both pneumatic and electronic types of controllers. The operator is responsible for recording the date, shift, equipment, ID, product and amount, indicating thermometer temperature, cleansing cycles, cut in and cut out temperatures, any connects for unusual circumstances, and his/her signature.

Flow Diversion Device (FDD)

Also called the flow diversion valve (FDV), it is located at the downstream end of the upward sloping holding tube. It is essentially a three-way valve, which, at temperatures greater than 72° C, opens to forward flow. This step requires power. At temperatures less than 72° C, the valve recloses to the normal position and diverts the milk back to the balance tank. It is important to note that the FDD operates on the measured temperature, not time, at the end of the holding period. There are two types of FDD:

1. *single stem* - an older valve system that has the disadvantage that it can't be cleaned in place.
2. *dual stem* - consists of 2 valves in series for additional fail safe systems. This FDD can be cleaned in place and is more suited for automation.

Flow Diversion Devices 17 KB

Vacuum Breaker

At the pasteurized product discharge is a vacuum breaker which breaks to atmospheric pressure. It must be located greater than 12 inches above the highest point of raw product in system. It ensures that nothing downstream is creating suction on the pasteurized side.

Auxiliary Equipment

Booster Pump

It is centrifugal "stuffing" pump which supplies raw milk to the raw regenerator for the balance tank. It must be used in conjunction with pressure differential controlling device and shall operate only when timing pump is operating, proper pressures are achieved in regenerator, and system is in forward flow.

Homogenizer

The homogenizer may be used as timing pump. It is a positive pressure pump; if not, then it cannot supplement flow. Free circulation from outlet to inlet is required and the speed of the homogenizer must be greater than the rate of flow of the timing pump.

MAGNETIC FLOW METER AND CENTRIFUGAL PUMP ARRANGEMENTS

Magnetic flow meters can be used to measure the flow rate. It is essentially a short piece of tubing (approximately 25 cm long) surrounded by a housing, inside of which are located coils that generate a magnetic field. When milk passes through the magnetic field, it causes a voltage to be induced, and the generated signal is directly proportional to velcoity. Application of the magnetic flow meter in the dairy industry has centered around its replacing the positive displacement timing pump as the metering device in HTST pasteurizing systems, where with certain products the timing pump rotors reportedly wear out in a relatvely short period of time. In operation, the electrical signal is sent by the magnetic flow meter to the flow controller, which determines what the actual flow is compared to the flow rate set by the operator. Since the magnetic flow meter continuously senses flow rate, it will signal the electronic controller if the actual flow exceeds the set flow rate for any reason. If the flow rate is exceeded for any reason, the flow diversion device is put into diverted flow. A significant difference from the normal HTST system (with timing pump) comes into focus at this point. This system can be operated at a flow rate greater than (residence time less than) the legal limit. However, it will be in diverted flow and never in forward flow.

Another magnetic flow meter based system with an AC variable frequency motor control drive on a centrifugal pump is also possible in lieu of a positive displacement metering pump on a HTST pasteurizer.This system does not use a control valve but rather the signal from the magnetic flow meter is transmitted to the AC variable frequency control to vary the speed of the centrifugal pump. The pump, then controls the flow rate of product through the system and its holding time in the holding tube.

Automated Public Health Controllers

These systems are used for time and temperature control of HTST systems. There are concerns that with sequential control, the critical control points (CCP's) are not monitored all the time; if during the sequence it got held up, the CCP's would not be monitored. With operator control, changes can be made to the program which might affect CCP's; the system is not easily sealed. No computer program can be written completely error free in large systems; as complexity increases, so too do errors.

This gives rise to a need for specific regulations or computer controlled CCP's of public health significance:

- *dedicated computer* - no other assignments, monitor all CCP's at least once/sec.
- not under control of any other computer system or override system, i.e., network.
- separate computer on each pasteurizer.
- I/O bus for outputs only, to other computers no inputs from other computers.
- on loss of power - public health computers should revert to fail safe position (e.g. divert).
- last state switches during power up must be fail safe position.
- programmes in ROM - tapes/disks not acceptable.
- inputs must be sealed, modem must be sealed, programme sealed.
- no operator override switches.
- proper calibration procedure during that printing - Public health computer must not leave public health control for > 1 sec and upon return must complete 1 full cycle before returning to printing.
- FDV position must be monitored and temperature in holding tube recorded during change in FDV position.
- download from ROM to RAM upon startup.
- integrated with CIP computer which can be programmed e.g., FDV, booster pump controllable by CIP computer when in CIP made only.

UHT PROCESSING

While pasteurization conditions effectively eliminate potential pathogenic microorganisms, it is not sufficient to inactivate the thermoresistant spores in milk. The term sterilization refers to the complete elimination of all microorganisms. The food industry uses the more realistic term "commercial sterilization"; a product is not necessarily free of all micro-organisms, but those that survive the sterilization process are unlikely to grow during storage and cause product spoilage.

In canning we need to ensure the "cold spot" has reached the desired temperature for the desired time. With most canned products, there is a low rate of heat penetration to the thermal centre. This leads to overprocessing of some portions, and damage to nutritional and sensory characteristics, especially near the walls of the container. This implies long processing times at lower temperatures.

Milk can be made commercially sterile by subjecting it to temperatures in excess of 100° C, and packaging it in air-tight containers. The milk may be packaged either before or after sterilization. The basis of UHT, or ultra-high temperature, is the sterilization of food before packaging, then filling into pre-sterilized containers in a sterile atmosphere. Milk that is processed in this way using temperatures exceeding 135° C, permits a decrease in the necessary holding time (to 2-5 s) enabling a continuous flow operation.

Some examples of food products processed with UHT are:

- *liquid products* - milk, juices, cream, yoghurt, wine, salad dressings
- *foods with discrete particles* - baby foods; tomato products; fruits and vegetables juices; soups
- *larger particles* - stews

Advantages of UHT

High Quality

The D and Z valves are higher for quality factors than microorganisms. The reduction in process time due to higher temperature (UHTST) and the minimal come-up and cool-down time leads to a higher quality product.

Long Shelf Life

Greater than six months, without refrigeration, can be expected.

Packaging Size

Processing conditions are independent of container size, thus allowing for the filling of large containers for food-service or sale to food manufacturers (aseptic fruit purees in stainless steel totes).

Cheaper Packaging

Both cost of package and storage and transportation costs; laminated packaging allows for use of extensive graphics

Difficulties with UHT

Sterility

Complexity of equipment and plant are needed to maintain sterile atmosphere between processing and packaging (packaging materials, pipework, tanks, pumps); higher skilled operators; sterility must be maintained through aseptic packaging

Particle Size

With larger particulates there is a danger of overcooking of surfaces and need to transport material - both limits particle size.

Equipment

There is a lack of equipment for particulate sterilization, due especially to settling of solids and thus overprocessing

Keeping Quality

Heat stable lipases or proteases can lead to flavour deterioration, age gelation of the milk over time - nothing lasts forever! There is also a more pronounced cooked flavour to UHT milk.

UHT Methods

There are two principal methods of UHT treatment:

1. Direct Heating system; and
2. Indirect Heating.

Direct Heating Systems

The product is heated by direct contact with steam of potable or culinary quality. The main advantage of direct heating is that the product is held at the elevated temperature for a shorter period of time. For a heat-sensitive product such as milk, this means less damage.

Indirect vs Direct Heating (17 KB)

There are two methods of direct heating:

1. injection;
2. infusion.

Injection

High pressure steam is injected into pre-heated liquid by a steam injector leading to a rapid rise in temperature. After holding, the product is flash-cooled in a vacuum to remove water equivalent to amount of condensed steam used. This method allows fast heating and cooling, and volatile removal, but is only suitable for some products. It is energy intensive and because the product comes in contact with hot equipment, there is potential for flavour damage.

Infusion

The liquid product stream is pumped through a distributing nozzle into a chamber of high pressure steam. This system is characterized by a large steam volume and a small product volume, distributed in a large surface area of product. Product temperature is accurately controlled via pressure. Additional holding time may be accomplished through the use of plate or tubular heat exchangers, followed by flash cooling in vacuum chamber. This method has several advantages:

- instantaneous heating and rapid cooling;
- no localized overheating or burn-on;
- suitable for low and higher viscosity products;

Indirect Heating Systems

The heating medium and product are not in direct contact, but separated by equipment contact surfaces. Several types of heat exchangers are applicable:

- plate;
- tubular; and
- scraped surface.

Plate Heat Exchangers

Similar to that used in HTST but operating pressures are limited by gaskets. Liquid velocities are low which could lead to uneven heating and burn-on. This method is economical in floor space, easily inspected, and allows for potential regeneration.

Tubular Heat Exchangers

There are several types:

- shell and tube;
- shell and coil;
- double tube;
- triple tube;

All of these tubular heat exchangers have fewer seals involved than with plates. This allows for higher pressures, thus higher flow rates and higher temperatures. The heating is more uniform but difficult to inspect.

Scraped Surface Heat Exchangers: The product flows through a jacketed tube, which contains the heating medium, and is scraped from the sides with a rotating knife. This method is suitable for viscous products and particulates (< 1 cm) such as fruit sauces, and can be adjusted for different products by changing configuration of rotor. There is a problem with larger particulates; the long process time for particulates would mean long holding sections which are impractical. This may lead to damaged solids and overprocessing of sauce.

PACKAGING FOR ASEPTIC PROCESSING

The most important point to remember is that it must be sterile! All handling of product post-process must be within the sterile environment.

There are five basic types of aseptic packaging lines:

1. *Fill and seal:* preformed containers made of thermoformed plastic, glass or metal are sterilized, filled in aseptic environment, and sealed.
2. *Form, fill and seal:* roll of material is sterilized, formed in sterile environment, filled, sealed e.g. tetrapak
3. *Erect, fill and seal:* using knocked-down blanks, erected, sterilized, filled, sealed. e.g. gable-top cartons, cambri-bloc
4. Thermoform, fill, sealed roll stock sterilized, thermoformed, filled, sealed aseptically. e.g. creamers, plastic soup cans
5. *Blow mold, fill, seal:* There are several different package forms that are used in aseptic UHT processing:
   - cans;
   - paperboard/plastic/foil/plastic laminates;
   - flexible pouches;
   - thermoformed plastic containers;
   - flow molded containers;
   - bag-in-box; and
   - bulk totes.

It is also worth mentioning that many products that are UHT heat treated are not aseptically packaged. This gives them the advantage of a longer shelf life at refrigeration temperatures compared to pasteurization, but it does not produce a shelf-stable product at ambient temperatures, due to the possibility of recontamination post-processing.

HOMOGENIZATION OF MILK AND MILK PRODUCTS

Introduction

Milk is an oil-in-water emulsion, with the fat globules dispersed in a continuous skimmilk phase. If raw milk were left to stand, however, the fat would rise and form a cream layer. Homogenization is a mechanical treatment of the fat globules in milk brought about by passing milk under high pressure through a tiny orifice, which results in a decrease in the average diameter and an increase in number and surface area, of the fat globules. The net result, from a practical view, is a much reduced tendency for creaming of fat globules. Three factors contribute to this enhanced stability of homogenized milk: a decrease in the mean diameter of the fat globules (a factor in Stokes Law), a decrease in the size distribution of the fat globules (causing the speed of rise to be similar for the majority of globules such that they don't tend to cluster during creaming), and an increase in density of the globules (bringing them closer to the continuous phase) oweing to the adsorption of a protein membrane. In addition, heat pasteurization breaks down the cryo-globulin complex, which tends to cluster fat globules causing them to rise.

HOMOGENIZATION MECHANISM

Auguste Gaulin's patent in 1899 consisted of a 3-piston pump in which product was forced through one or more hair like tubes under pressure. It was discovered that the size of fat globules produced were 500 to 600 times smaller than tubes. There have been over 100 patents since, all designed to produce smaller average particle size with expenditure of as little energy as possible. The homogenizer consists of a 3-cylinder positive piston pump (operates similar to car engine) and homogenizing valve. The pump is turned by electric motor through connecting rods and crankshaft.

To understand the mechanism, consider a conventional homogenizing valve processing an emulsion such as milk at a flow rate of 20,000 l/h at 14 MPa (2100 psig). As it first enters the valve, liquid velocity is about 4 to 6 m/s. It then moves into the gap between the valve and the valve seat and its velocity is increased to 120 metre/sec in about 0.2 millisec. The liquid then moves across the face of the valve seat (the land) and exits in about 50 microsec. The homogenization phenomena is completed before the fluid leaves the area between the valve and the seat, and therefore emulsification is initiated and completed in less than 50 microsec. The whole process occurs between two pieces of steel in a steel valve assembly. The product may then pass through a second stage valve similar to the first stage. While most of the fat globule reduction takes place in the first stage, there is a tendency for clumping or clustering of the reduced fat globules. The second stage valve permits the separation of those clusters into individual fat globules.

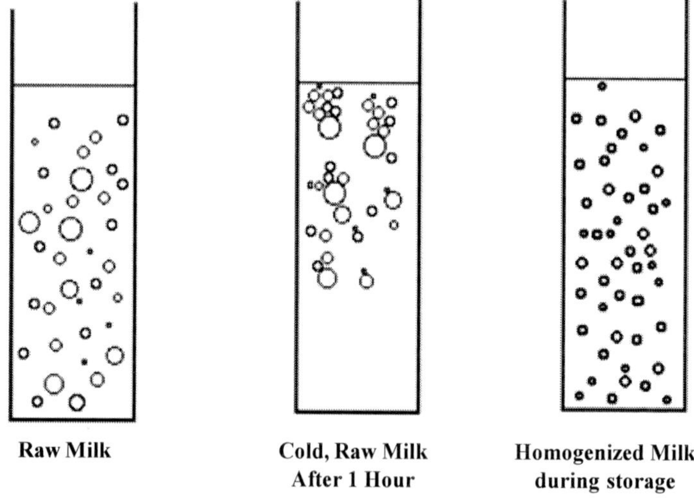

Fig. 6.1: **Homogenization of Milk**

It is most likely that a combination of two theories, turbulence and cavitation, explains the reduction in size of the fat globules during the homogenization process.

Turbulence

Energy, dissipating in the liquid going through the homogenizer valve, generates intense turbulent eddies of the same size as the average globule diameter. Globules are thus torn apart by these eddie currents reducing their average size.

Cavitation

Considerable pressure drop with charge of velocity of fluid. Liquid cavitates because its vapor pressure is attained. Cavitation generates further eddies that would produce disruption of the fat globules.

The high velocity gives liquid a high kinetic energy which is disrupted in a very short period of time. Increased pressure increases velocity. Dissipation of this energy leads to a high energy density (energy per volume and time). Resulting diameter is a function of energy density.

In summary, the homogenization variables are:

- type of valve;
- pressure;
- single or two-stage;
- fat content;
- surfactant type and content;
- viscosity; and

temperature.

Also to be considered are the droplet diameter (the smaller, the more difficult to disrupt), and the log diameter which decreases linearly with log P and levels off at high pressures.

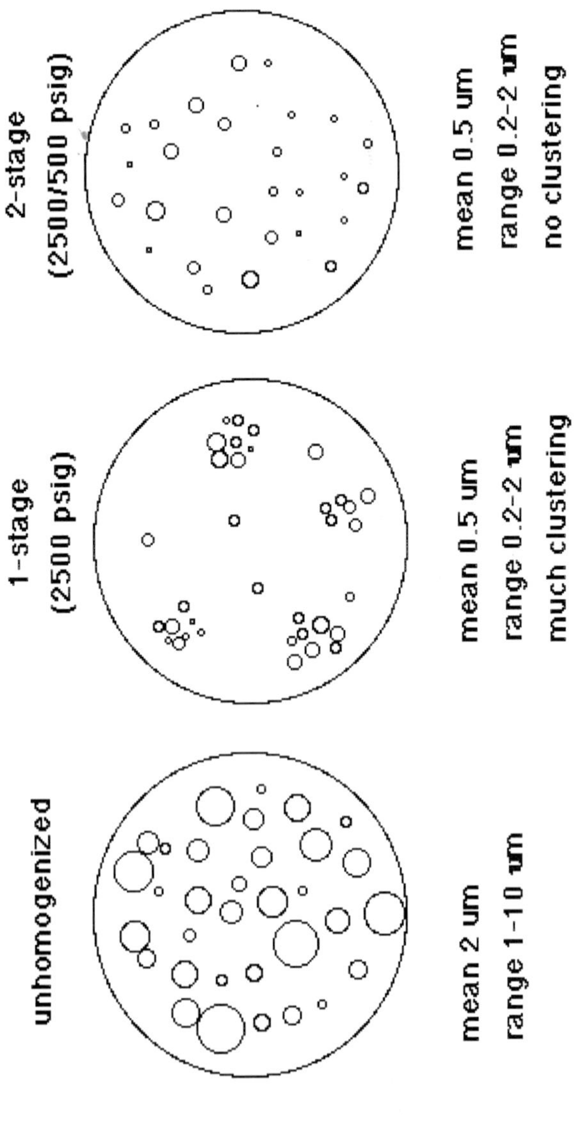

Fig. 6.2: **The Effects of 2-stage Homogenization on Fat Globule Size Distribution as Seen Under the Light Microscope**

Effect of Homogenization

Fat Globule

	No Homogenization	15 MPa (2500 psig)
Av. diam. (μ m)	3.3	0.4
Max. diam. (μ m)	10	2
Surf. area (m^2/ml of milk)	0.08	
Number of globules (μ m^{-3})	0.02	12

Surface Layer

The milk fat globule has a native membrane, picked up at the time of secretion, made of amphiphilic molecules with both hydrophilic and hydrophobic sections. This membrane lowers the interfacial tension resulting in a more stable emulsion. During homogenization, there is a tremendous increase in surface area and the native milk fat globule membrane (MFGM) is lost. However, there are many amphiphilic molecules present from the milk plasma that readily adsorb: casein micelles (partly spread) and whey proteins. The interfacial tension of raw milk is 1-2 mN/m, immediately after homogenization it is unstable at 15 mN/m, and shortly becomes stable (3-4 mN/m) as a result of the adsorption of protein. The transport of proteins is not by diffusion but mainly by convection. Rapid coverage is achieved in less than 10 sc but is subject to some rearrangement.

Surface excess is a measure of how much protein is adsorbed; for example 10 mg/m2 translates to a thickness of adsorbed layer of approximately 15 nm

Membrane Processing

Membrane processing is a technique that permits concentration and separation without the use of heat. Particles are separated on the basis of their molecular size and shape with the use of pressure and specially designed semi-permeable membranes. There are some fairly new developments in terms of commercial reality and is gaining readily in its applications:

- proteins can be separated in whey for the production of whey protein concentrate (WPC)
- milk can be concentrated prior to cheesemaking at the farm level
- apple juice and wine can be clarified
- waste treatment and product recovery is possible in edible oil, fat, potato, and fish processing
- fermentation broths can be clarified and separated
- whole egg and egg white ultrafiltration as a preconcentration prior to spray drying

The following topics will be covered in this section:

Principle of Operation

Types of Membrane Processing

Reverse Osmosis

Ultrafiltration

Microfiltration

Hardware Design

Electrodialysis

Ion Exchange

Principle of Operation

When a solution and water are separated by a semi-permeable membrane, the water will move into the solution to equilibrate the system. This is known as osmotic pressure If a mechanical force is applied to exceed the osmotic pressure (up to 700 psi), the water is forced to move down the concentration gradient i.e. from low to high concentration. Permeate designates the liquid passing through the membrane, and retentate (concentrate) designates the fraction not passing through the membrane.

Membrane Processing

Reverse Osmosis

Reverse osmosis (RO) designates a membrane separation process, driven by a pressure gradient, in which the membrane separates the solvent (generally water) from other components of a solution. The membrane configuration is usually cross-flow. With reverse osmosis, the membrane pore size is very small allowing only small amounts of very low molecular weight solutes to pass through the membranes. It is a concentration process using a 100 MW cutoff, 700 psig, temperatures less than 40°C with cellulose acetate membranes and 70-80°C with composite membranes. (Fig. 6.3).

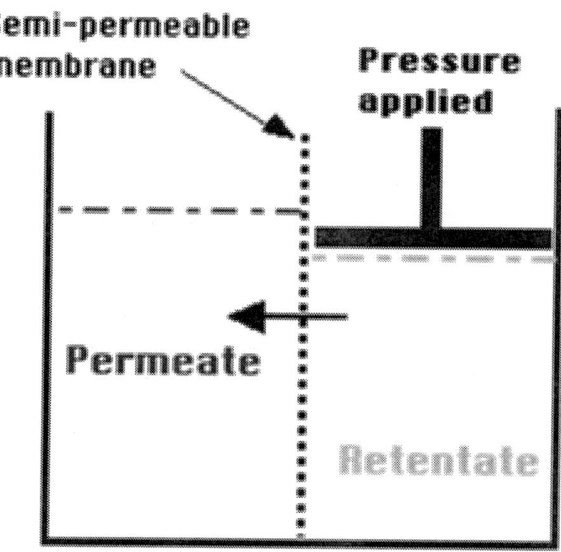

Fig 6.3: **Membrane Processing**

Hyperfiltration is the same as RO

Ultrafiltration

Ultrafiltration (UF) designates a membrane separation process, driven by a pressure gradient, in which the membrane fractionates components of a liquid as a function of their solvated size and structure. The membrane configuration is usually cross-flow. In UF, the membrane pore size is larger allowing some components to pass through the pores with the water. It is a separation/fractionation process using a 10,000 MW cutoff, 40 psig, and temperatures of 50-60°C with polysulfone membranes. In UF milk, lactose and minerals pass in a 50% separation ratio; for example, in the retentate would be 100% of fat, 100% of protein, 50% of lactose, and 50% of free minerals.

Diafiltration is a specialized type of ultrafiltration process in which the retentate is diluted with water and re-ultrafiltered, to reduce the concentration of soluble permeate components and increase further the concentration of retained components.

Microfiltration

Microfiltration (MF) designates a membrane separation process similar to UF but with even larger membrane pore size allowing particles in the range of 0.2 to 2 micrometers to pass through. The pressure used is generally lower than that of UF process. The membrane configuration is usually cross-flow. MF is used in the dairy industry for making low-heat sterile milk as proteins may pass through but bacteria do not. Please click above link for a schematic diagram of these membrane processes.

Hardware Design

Open Tubular

Tubes of membrane with a diameter of 1/2 to 1 inch and length to 12 ft. are encased in reinforced fibreglass or enclosed inside a rigid PVC or stainless steel shell. As the feed solution flows through the membrane core, the permeate passes through the membrane and is collected in the tubular housing. Imagine 12 ft long straws!

Hollow Fibre

Similar to open tubular, but the cartridges contain several hundred very small (1 mm diam) hollow membrane tubes or fibres. As the feed solution flows through the open cores of the fibres, the permeate is collected in the cartridge area surrounding the fibres.

Plate and Frame

This system is set up like a plate heat exchanger with the retentate on one side and the permeate on the other. The permeate is collected through a central collection tube.

Spiral Wound

This design tries to maximize surface area in a minimum amount of space. It consists of consecutive layers of large membrane and support material in an envelope type design rolled up around a perforated steel tube.

Electrodialysis

Electrodialysis is used for demineralization of milk products and whey for infant formula and special dietary products. Also used for desalination of water.

Principles of Operation

Under the influence of an electric field, ions move in an aqueous solution. The ionic mobility is directly proportioned to specific conductivity and inversely proportioned to number of molecules in solution. ~3-6 × 102 mm/sec.

Charged ions can be removed from a solution by synthetic polymer membranes containing ion exchange groups. Anion exchange membranes carry cationic groups which repel cations and are permeable to anions, and cation exchange membranes contain anionic groups and are permeable only to cations.

Electrodialysis membranes are comprised of polymer chains - styrene-divinyl benzene made anionic with quaternary ammonium groups and made cationic with sulphonic groups. 1-2V is then applied across each pair of membranes.

Electrodialysis Process

Amion and cation exchange membranes are arranged alternately in parallel between an anode and a cathode (see schematic diagram). The distance between the membranes is 1mm or less. A plate and frame arrangement similar to a plate heat exchanger or a plate filter is used. The solution to be demineralized flows through gaps between the two types of membranes. Each type of membrane is permeable to only one type of ion. Thus, the anions leave the gap in the direction of the anode and cations leave in the direction of the cathode. Both are then taken up by a concentrating stream.

Problems

Concentration polarization. Deposits on membrane surfaces, e.g. proteins - pH control is important. Prior concentration of whey, to 20% TS, is necessary before electrodialysis.

Ion Exchange

Ion exchange is not a membrane process but I have included it here anyway because it is used for product of protein isolates of higher concentration than obtainable by membrane concentration.

Fractionation may also be accomplished using ion exchange processing. It relies on inert resins (cellulose or silica based) that can adsorb charged particles at either end of the pH scale. The design can be a batch type, stirred tank or continuous column. The column is more suitable for selective fractionation. Whey protein isolate (WPI), with a 95% protein content, can be produced by this method. Following adsorption and draining of the deproteined whey, the pH or charge properties are altered and proteins are eluted. Protein is recovered from the dilute stream through UF and drying. Selective resins may be used for fractionated protein products or enriched in fraction allow tailoring of ingredients.

Evaporation and Dehydration

The removal of water from foods provides microbiological stability, reduces deteriorative chemical reactions, and reduces transportation and storage costs. Both evaporation and dehydration are methods used in the dairy industry for this purpose. The following topics will be addressed here:

- Evaporation
- Principle of Operation
- Evaporator designs
    - Batch Pan
    - Rising Film
    - Falling Film

- Multiple Effect Evaporators
  - Thermo compression
  - Mechanical Vapour Recompression
- Dehydration
- Spray Drying Process
- Powder Recovery
- Bag filters
- Cyclone collector
- Wet scrubber
- Two and Three Stage Spray Driers
- Principles of Fluid Beds
- Process
- Agglomeration and Instantizing

Evaporation

Evaporation refers to the process of heating liquid to the boiling point to remove water as vapour. Because milk is heat sensitive, heat damage can be minimized by evaporation under vacuum to reduce the boiling point. The basic components of this process consist of:

- heat-exchanger
- vacuum
- vapour separator
- condenser

The heat exchanger is enclosed in a large chamber and transfers heat from the heating medium, usually low pressure steam, to the product usually via indirect contact surfaces. The vacuum keeps the product temperature low and the difference in temperatures high. The vapour separator removes entrained solids from the vapours, channelling solids back to the heat exchanger and the vapours out to the condenser. It is sometimes a part of the actual heat exchanger, especially in older vacuum pans, but more likely a separate unit in newer installations. The condenser condenses the vapours from inside the heat exchanger and may act as the vacuum source.

Principle of Operation

The driving force for heat transfer is the difference in temperature between the steam in the coils and the product in the pan. The steam is produced in large boilers, generally tube and chest heat exchangers. The steam temperature is a function of the steam pressure. Water boils at 100°C at 1 atm., but at other pressures the boiling point changes. At its boiling point, the steam condenses in the coils and gives up its latent heat. If the steam temperature is too high, burn-on/fouling increases so there are limits to how high steam temperatures can go. The product is also at its boiling point. The boiling point can be elevated with an increase in solute concentration. This boiling point elevation works on the same principles as freezing point depression.

Evaporator Designs

Types of single effect evaporators:

- Batch Pan
- Rising film
- Falling film
- Plate evaporators
- Scraped surface

Batch pan evaporators are the simplest and oldest. They consist of spherical shaped, steam jacketed vessels. The heat transfer per unit volume is small requiring long residence times. The heating is due only to natural convection, therefore, the heat transfer characteristics are poor. Batch plants are of historical significance; modern evaporation plants are far-removed from this basic idea. The vapours are a tremendous source of low pressure steam and must be reused.

Rising film evaporators consist of a heat exchanger isolated from the vapour separator. The heat exchanger, or calandria, consists of 10 to 15 metre long tubes in a tube chest which is heated with steam. The liquid rises by percolation from the vapours formed near the bottom of the heating tubes. The thin liquid film moves rapidly upwards. The product may be recycled if necessary to arrive at the desired final concentration. This development of this type of modern evaporator has given way to the falling film evaporator.

The falling film evaporators are the most widely used in the food industry. They are similar in components to the rising film type except that the thin liquid film moves downward under gravity in the tubes. A uniform film distribution at the feed inlet is much more difficult to obtain. This is the reason why this development came slowly and it is only within the last decade that falling film has superceded all other designs. Specially designed nozzles or spray distributors at the feed inlet permit it to handle more viscous products. The residence time is 20-30 sec. as opposed to 3-4 min. in the rising film type. The vapour separator is at the bottom which decreases the product hold-up during shut down. The tubes are 8-12 metres long and 30-50 mm in diameter.

Multiple Effect Evaporators

Two or more evaporator units can be run in sequence to produce a multiple effect evaporator. Each effect would consist a heat transfer surface, a vapour separator, as well as a vacuum source and a condenser. The vapours from the preceding effect are used as the heat source in the next effect. There are two advantages to multiple effect evaporators:

- *economy* - they evaporate more water per kg steam by re-using vapours as heat sources in subsequent effects;
- *improve heat transfer* - due to the viscous effects of the products as they become more concentrated.

Each effect operates at a lower pressure and temperature than the effect preceding it so as to maintain a temperature difference and continue the evaporation procedure. The vapours are removed from the preceding effect at the boiling temperature of the product at that effect so that no temperature difference would exist if the vacuum were not increased. The operating costs of evaporation are relative to the number of effects and the temperature at which they operate. The boiling milk creates vapours which can be recompressed for high steam econonmy. This can be done by adding energy to the vapour in the form of a steam jet, thermo compression or by a mechanical compressor, mechanical vapour recompression.

Thermo Compression (TC)

Involves the use of a steam-jet booster to recompress part of the exit vapours from the first effect. Through recompression, the pressure and temperature of the vapours are increased. As the vapours exit from the first effect, they are mixed with very high pressure steam. The steam entering the first effect calandria is at slightly less pressure than the supply steam. There is usually more vapours from the first effect than the second effect can use; usually only the first effect is coupled with multiple effect evaporators.

Mechanical Vapour Recompression (MVR)

Whereas only part of the vapour is recompressed using TC, all the vapour is recompressed in an MVR evaporator. Vapours are mechanically compressed by radial compressors or simple fans using electrical energy.

There are several variations; in single effect, all the vapours are recompressed therefore no condensing water is needed; in multiple effect, can have MVR on first effect, followed by two or more traditional effects; or can recompress vapours from all effects.

Dehydration

Dehydration refers to the nearly complete removal of water from foods to a level of less than five per cent. Although there are many types of driers, spray driers are the most widely used type of air convection drier. It turns out more tonnage of dehydrated products than all other types of driers combined. It is limited to food that can be atomized, i.e. liquids, low viscosity pastes, and purees. Drying takes place within a matter of seconds at temperatures approximately 200° C. Evaporative cooling maintains low product temperatures, however, prompt removal of the product is still necessary.

Spray Drying—Process Summary

The liquid food is generally preconcentrated by evaporation to economically reduce the water content. The concentrate is then introduced as a fine spray or mist into a tower or chamber with heated air. As the small droplets make intimate contact with the heated air, they flash off their moisture, become small particles, and drop to the bottom of the tower and are removed. The advantages of spray drying include a low heat and short time combination which leads to a better quality product.

Spray Dryer 17 KB

Principal components include:

- a high pressure pump for introducing liquid into the tower;
- a device for atomizing the feed stream;
- a heated air source with blower;
- a secondary collection vessel for removing the dried food from the airstream;
- means for exhausting the moist air.
- usually includes a preconcentration step i.e. MVR evaporation

Atomizing devices are the distinguishing characteristic of spray drying. They provide a large surface area for exposure to drying forces:

1 litre = 12 billion particles = >300 ft2 ($30m^2$)

The exit air temperature is an important parameter to monitor because it responds readily to changes in the process and reflects the quality of the product. Generally, we want it high enough to yield desired moisture without heat damage. There are two controls that may be used to adjust the exit air temperature:

- altering feed flow rate
- altering inlet temperature

If heat damage occurs before the product is dried, the particle size must be reduced; smaller particle dries faster, therefore, less heat damage. This can be accomplished in three ways:

- smaller orifice
- increase atomizing pressure.
- reduce viscosity - by increasing feed temperature or reducing solids

Powder Recovery

It is essential for both economic and environmental reasons that as much powder as possible be recovered from the air stream. Three systems are available, however wet scrubbers usually act as a secondary collection system following a cyclone.

Bag Filters

Bag filters are very efficient (99.9%), but not as popular due to labor costs, sanitation, and possible heat damage because of the long residence time. They are not recommended in the case of handling high moisture loads or hygroscopic particles.

Cyclone Collector

Cyclones are not as efficient (99.5%) as bag filtres but several can be placed in series. Air enters at tangent at high velocity into a cylinder or cone which has a much larger cross section. Air velocity is decreased in the cone permitting settling of solids by gravity. Centrifugal force is important in removing particles from the air stream. High air velocity is needed to separate small diameter and light materials from air; velocities may approach 100 ft/sec (70 MPH). Higher centrifugal force can be obtained by using small diameter cyclones, several of which may be placed in parallel; losses may range from 0.5-2%. A rotary airlock is used to remove powder from the cyclone. (An example of a rotary airlock is a revolving door at a hotel lobby which is intended to break the outside and inside environments).

Wet Scrubber

Wet scrubbers are the most economical outlet air cleaner. The principle of a wet scrubber is to dissolve any dust powder left in the airstream into either water or the feed stream by spraying the wash stream through the air. This also recovers heat from the exiting air and evaporates some of the water in the feed stream (if used as the wash water).

Wet Scrubbers

Wet scrubbers not only recover most of what would be lost product, but also recover approximately 90% of the potential drying energy normally lost in exit air. The exit air picks up moisture which increases evaporative capacity by 8% (concentration of feed). Cyclone separators are probably the best primary powder separator system because they are hygienic, easy to operate, and versatile, however, high losses may occur. Wet scrubbers are designed for a secondary air cleaning system in conjunction with the cyclone. Either feed stream or water can be used as scrubbing liquor. Also, there are heat recovery systems available.

Air is blown up through a wire mesh belt on porous plate that supports and conveys the product. A slight vibration motion is imparted to the food particles. When the air velocity is increased to the point where it just exceeds the velocity of free fall (gravity) of the particles, fluidization occurs. The dancing/boiling motion subdivides the product and provides intimate contact of each particle with the air, but keeps clusters from forming.

With products that are particularly difficult to fluidize, a vibrating motion of the drier itself is used to aid fluidization;

it is called vibro-fluidizer which is on springs. The fluidized solid particles then behave in an analogous manner to a liquid., i.e. they can be conveyed. Air velocities will vary with particle size and density, but are in the range of 0.3 - 0.75 m/s. They can be used not only for drying but also for cooling. If the velocity is too high, the particles will be carried away in the gas stream, therefore, gravitational forces need to be only slightly exceeded.

Two and Three Stage Drying Processes

In standard, single stage spray drying, the rate of evaporation is particularly high in the first part of the process, and it gradually decreases because of the falling moisture content of the particle surfaces. In order to complete the drying in one stage, a relatively high outlet temperature is required during the final drying phase. Of course the outlet temperature is reflective of the particle temperature and thus heat damage.

Consequently, the two stage drying process was introduced which proved to be superior to the traditional single stage drying in terms of product quality and cost of production.

The two-stage drier consists of a spray drier with an external vibrating fluid bed placed below the drying chamber. The product can be removed from the drying chamber with a higher moisture content, and the final drying takes place in the external fluid bed where the residence time of the product is longer and the temperature of the drying air lower than in the spray dryer.

This principle forms the basis of the development of the three stage drier. The second stage is a fluid bed built into the cone of the spray drying chamber. Thus it is possible to achieve an even higher moisture content in the first drying stage and a lower outlet air temperature from the spray drier. This fluid bed is called the integrated fluid bed. The inlet air temperature can be raised resulting in a larger temperature difference and improved efficiency in the drying process. The exhaust heat from the chamber is used to preheat the feed stream. The third stage is again the external fluid bed, which can be static or vibrating, for final drying and/or cooling the powder. The results are as follows:

- higher quality powders with much better rehydrating properties directly from the drier;
- lower energy consumption;
- increased range of products which can be spray dried i.e., non density, non hygroscopic;
- smaller space requirements.

Agglomerating and Instantizing

These processes have allowed the manufacturing of milk powders with better reconstitution properties, such as instantized skim milk powder.

Agglomeration Mechanism

Powder is wetted with water or steam. The surface must be uniformly wetted but not excessively. The powder is held wet over a selected period of time to give moisture stability to the clusters which have formed. The clusters are dried to the desired moisture content and then cooled (e.g., fluid bed). Dried clusters are screened and sized to reduce excessively large particles and remove excessively small ones. The agglomeration process causes an increase in the amount of air incorporated between powder particles. More incorporated air is replaced with more water when the powder is reconstituted, which immediately wet the powder particles.

AGGLOMERATING TECHNIQUES

Rewet Methods

This method uses powder as feed stock. An example is the ARCS Instantizer. Humidified air moistens powder, which causes it to cluster. It is re-dried and wetted. The clustered powder is then exposed to heated, filtered, high-velocity air. The dried clusters are then exposed to cooled air on a vibrating belt. It is then sized (pelleted to uniform size) and the fines are removed.

Straight Thru Process

A multi-stage drying process produces powders with much better solubility characteristics similar to instantized powder. This method uses a low outlet temperature which allows higher moisture in powder as it is taken from spray drier with excess moisture removed in the fluid bed. The powder fines are reintroduced to the atomizing cloud in the drying chamber.

Production and Utilization of Steam and Refrigeration

Steam Production and Utilization

Understanding steam

Figure 6.4 below helps to explain the various principles involved in the thermodynamics of steam. It shows the relationship between temperature and enthalpy (energy or heat content) of water as it passes through its phase change.

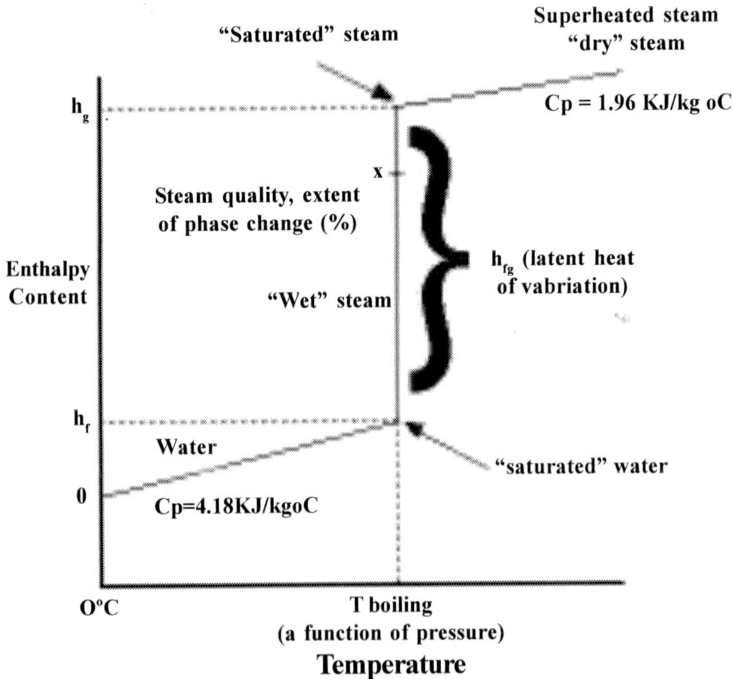

Fig. 6.4: Thermodynamics of steam

The reference point for enthalpy of water and steam is 0°C, at which point an enthalpy value of 0 kJ/kg is given to it (but of course water at 0o has alot of energy in it, which is given up as it freezes - it's not until 0K, absolute zero, when it truely has no enthalpy!). As we increase the temperature of water, its enthalpy increases by 4.18 kJ/kg °C until we hit its boiling point (which is a function of its pressure - the boiling point of water is 100° C ONLY at 1 atm. pressure). At this point, a large input of enthalpy causes no temperature change but a phase change, latent heat is added and steam is produced. Once all the water has vaporized, the temperature again increases with the addition of heat (sensible heat of the vapour).

Steam Production and Distribution

Steam is produced in large tube and chest heat exchangers, called water tube boilers if the water is in the tubes, surrounded by the flame, or fire tube boilers if the opposite is true. The pressure inside a boiler is usually high, 300-800 kPa. The steam temperature is a function of this pressure. The steam, usually saturated or of very high quality, is then distributed to the heat exchanger where it is to be used, and it provides heat by condensing back to water (called condensate) and giving up its latent heat. The temperature desired at the heat exchanger can be adjusted by a pressure reducing valve, which lowers the pressure to that corresponding to the desired temperature. After the steam condenses in the heat exchanger, it passes through a steam trap (which only allows water to pass through and hence holds the steam in the heat exchanger) and then the condensate (hot water) is returned to the boiler so it can be reused. The following image is a schematic of a steam production and distribution cycle.

Refrigeration

The following image is a schematic of a refrigeration cycle. It is described in detail below, so you may want to go back and forth between the diagram and the description.

Refrigeration Cycle

Mechanical refrigerators have four basic elements: an evaporator, a compressor, a condenser, and a refrigerant flow control (expansion valve). A refrigerant circulates among the four elements changing from liquid to gas and back to liquid.

In the evaporator, the liquid refrigerant evaporates (boils) under reduced pressure and in doing so absorbs latent heat of vaporization and cools the surroundings. The evaporator is at the lowest temperature in the system and heat flows to it. This heat is used to vaporize the refrigerant. The temperature at which this occurs is a function of the pressure on the refrigerant: for example if ammonia is the refrigerant, at -18°C the ammonia pressure required is 1.1 kg/sq. cm. The part of the process described thus far is the useful part of the refrigeration cycle; the remainder of the process is necessary only so that the refrigerant may be returned to the evaporator to continue the cycle.

The refrigerant vapour is sucked into a compressor, a pump that increases the pressure and then exhausts it at a higher pressure to the condenser. For ammonia, this is approx. 10 kg/sq. cm. To complete the cycle, the refrigerant must be condensed back to liquid and in doing this it gives up its latent heat of vaporization to some cooling medium such as water or air. The condensing temperature of ammonia is 29°C, so that cooling water at about 21°C could be used. In home refrigerators, the compressed gas (not ammonia) is sent through the pipes at the back, which are cooled by circulating air around them. Often fins are added to these tubes to increase the cooling area. The gas had to be compressed so that it could be condensed at these higher temperatures, using free cooling from water or air.

The refrigerant is now ready to enter the evaporator to be used again. It passes through an expansion valve to enter into the region of lower pressure, which causes it to boil and absorb more heat from the load. By adjusting the high and low pressures, the condensing and evaporating temperatures can be adjusted as required.

7

Dairy Products

FLUID MILK PROCESSING

The production of beverage milks combines the unit operations of clarification, separation (for the production of lower fat milks), pasteurization, and homogenization. The process is simple, as indicated in the flow chart. While the fat content of most raw milk is 4% or higher, the fat content in most beverage milks has been reduced to 3.4%. Lower fat alternatives, such as 2% fat, 1% fat, or skim milk (<0.1% fat) or also available in most markets. These products are either produced by partially skimming the whole milk, or by completely skimming it and then adding an appropriate amount of cream back to achieve the desired final fat content.

Vitamins may be added to both full fat and reduced fat milks. Vitamins A and D (the fat soluble ones) are often supplemented in the form of a water soluble emulsion to offset that quantity lost in the fat separation process.

Creams

During the separation of whole milk, two streams are produced: the fat-depleted stream, which produces the beverage milks as described above or skim milk for evaporation and possibly for subsequent drying, and the fat-rich stream, the cream. This usually comes off the separator with fat

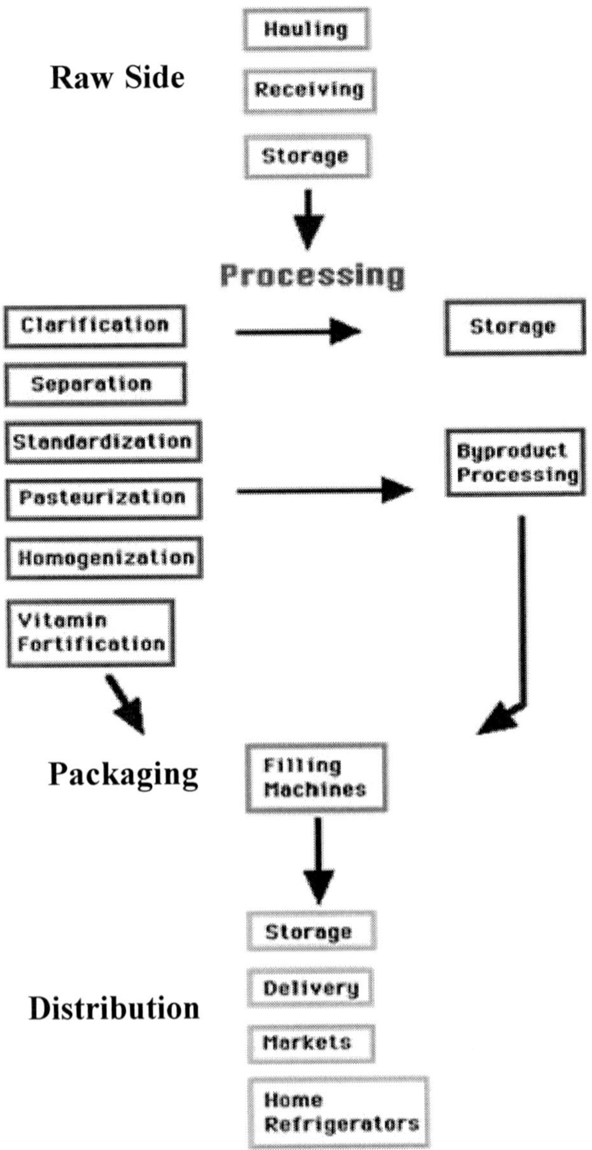

Fig. 7.1: Milk Processing Scheme

contents in the 35-45% range. Cream is used for further processing in the dairy industry for the production of ice-cream or butter, or can be sold to other food processing industries. These industrial products normally have higher fat contents than creams for retail sale, normally in the range of 45-50% fat. A product known as "plastic" cream can be produced from certain types of milk separators. This product has a fat content approaching 80% fat, but it remains as an oil-in-water emulsion (the fat is still in the form of globules and the skim milk is the continuous phase of the emulsion), unlike butter which also has a fat content of 80% but which has been churned so that the fat occupies the continuous phase and the skim milk is dispersed throughout in the form of tiny droplets (a water-in-oil emulsion).

For retail cream products, the fat is normally standardized to 35% (heavy cream for whipping), 18% or 10% (cream for coffee or cereal). Higher fat creams have also been produced for retail sale, a product known as double cream has a fat content of 55% and is quite thick. Creams for packaging and sale in the retail market must be pasteurized to ensure freedom from pathogenic bacteria. Whipping cream is not normally homogenized, as the high fat content will lead to extensive fat globule aggregation and clustering, which leads to excessive viscosity and a loss of whipping ability. This phenomena has been used, however, to produce a spoonable cream product to be used as a dessert topping. Lower fat creams (10% or 18%) can be homogenized, usually at lower pressure than whole milk.

RECOMBINED MILK

Beverage milks can also be prepared by recombining skim milk powder and butter with water. This is often done in countries where there is not enough milk production to meet the demand for beverage milk consumption. The concept is simple. Skim milk powder is dispersed in water and allowed to hydrate. Butter is then emulsified into this mixture by either blending melted butter into the liquid mixture while hot, or by dispersing solid butter into the liquid through a high shear blender device. In some cases, a non-dairy fat source may also be used. The recombined milk product is then pasteurized, homogenized and packaged as in regular milk production. The final composition is similar to that of whole milk, approximately 9% milk solids-not-fat, and either 2% or 3.4% fat. The water source must be of excellent quality. The milk powder used for recombining must be of high quality and good flavour. Care must be taken to ensure adequate blending of the ingredients to prevent aggregation or lumping of the powder. Its dispersal in water is the key to success.

Chocolate Milk

An industry standard for the production of chocolate milk consists of:

- 93% milk
- 6.3% sugar
- 0.65% cocoa powder
- 0.05% carrageenan

The final product is usually standardized to either two per cent fat or one per cent fat (meaning, 2.15% or 1.1% fat in the milk before addition of other ingredients). The sugar, cocoa powder and carrageenan are dry blended, and added to cold milk with vigourous agitation, and then pasteurized.

CONCENTRATED AND DRIED DAIRY PRODUCTS

Fluid milk contains approximately 88% water. Concentrated milk products are obtained through partial water removal. Dried dairy products have even greater amounts of water removed to usually less than 4%. The benefits of both these processes include an increased shelf-life, convenience, product flexibility, decreased transportation costs, and storage.

The following products will be discussed here:

- Concentrated Dairy Products
  - Evaporated Skim or Whole Milk
  - Sweetened Condensed Milk

  - Condensed Buttermilk
  - Condensed Whey
- Dried Dairy Products
  - Milk Powder
  - Whey Powder
  - Whey Protein Concentrates

The principles of evaporation and dehydration can be found in the Dairy Processing section.

Concentrated Dairy Products

Evaporated Skim or Whole Milk

After the raw milk is clarified and standardized, it is given a pre-heating treatment of 93-100° C for 10 to 25 min or 115-128° C for 1 to 6 min. There are several benefits to this treatment:

- increases the concentrated milk stability during sterilization; decreases the chance of coagulation taking place during storage;
- decreases the initial microbial load;
- modifies the viscosity of the final product;
- milk enters the evaporator already hot.

Milk is then concentrated at low temperatures by vacuum evaporation. This process is based on the physical law that the boiling point of a liquid is lowered when the liquid is exposed to a pressure below atmospheric pressure. In this case, the boiling point is lowered to approximately 40-45° C. This results in little to no cooked flavour. The milk is concentrated to 30-40% total solids.

The evaporated milk is then homogenized to improve the milkfat emulsion stability. There are other benefits particular to this type of product:

- increased white colour

- increased viscosity
- decreased coagulation ability

A second standardization is done at this time to ensure the proper salt balance is present. The ability of milk to withstand intensive heat treatment depends to a great degree on its salt balance.

The product at this point is quite perishable. The fat is easily oxidized and the microbial load, although decreased, is still a threat. The evaporated milk at this stage is often shipped by the tanker for use in other products.

In order to extend the shelf life, evaporated milk can be packaged in cans and then sterilized in an autoclave. Continuous flow sterilization followed by packaging under aseptic conditions is also done. While the sterilization process produces a light brown colouration, the product can be successfully stored for up to a year.

Sweetened Condensed Milk

Where evaporated milk uses sterilization to extend its shelf-life, sweetened condensed milk has an extended shelf-life due to the addition of sugar. Sucrose, in the form of crystals or solution, increases the osmotic pressure of the liquid. This in turn, prevents the growth of microorganisms.

The only real heat treatment (85-90° C for several seconds) this product recieves is after the raw milk has been clarified and standardized. The benefits of this treatment include totally destroying osmophilic and thermophilic microorganisms, inactivating lipases and proteases, decreases fat separation and inhibits oxidative changes. Unfortunately it also affects the final product viscosity and may promote the defect age gelation.

The milk is evaporated in a manner similar to the evaporated milk. Although sugar may be added before evaporation, post evaporation addition is recommended to avoid undesirable viscosity changes during storage. Enough sugar is added so that the final concentration of sugar is approximately 45 per cent.

The sweetened evaporated milk is then cooled and lactose crystallization is induced. The milk is inoculated, or seeded, with powdered lactose crystals, then rapidly cooled while being agitated. The lactose can crystalize without the seeding but there is the danger of forming crystals that are too large. This would result in a texture defect similar in ice cream called sandiness, which affects the mouthfeel. By seeding, the number of crystals increases and the size of those crystals decreases.

The product is packaged in smaller containers, such as cans, for retail sales and bulk containers for industrial sales.

Condensed Buttermilk

Buttermilk is a by-product of the butter industry. It can be evaporated on its own or it can be blended with skimmilk and dried to produce skimmilk powder. This blended product may oxidise readily due to the higher fat content. Condensed buttermilk is perishable and, therefore, the supply must be fresh and it must be stored cool.

Condensed Whey

In the process of cheesemaking, there is alot of whey that needs to be disposed of. One of the ways of utilizing cheesewhey is to condense it. The whey contains fat, lactose, ß -lactoglobulin, alpha-lactalbumin, and water. The fat is generally removed by centrifugation and churned as whey cream or used in ice cream. Evaporation is the first step in producing whey powder.

Dried Dairy Products

Milk Powder

Milk used in the production of milk powders is first clarified, standardized and then given a heat treatment. This heat treatment is usually more severe than that required for pasteurization. Besides destroying all the pathogenic and most of the spoilage microorganisms, it also inactivates the enzyme lipase which could cause lipolysis during storage. The milk is then evaporated prior to drying for the following reasons:

- less occluded air and longer shelf life for the powder;

- viscosity increase leads to larger powder particles;
- less energy required to remove part of water by evaporation; more economical.

Homogenization may be applied to decrease the free fat content. Spray drying is the most used method for producing milk powders. After drying, the powder must be packaged in containers able to provide protection from moisture, air, light, etc. Whole milk powder can then be stored for long periods (up to about 6 months) of time at ambient temperatures.

Instant milk powder is produced by partially rehydrating the dried milk powder particles causing them to become sticky and agglomerate. The water is then removed by drying resulting in an increased amount of air incorporated between the powder particles.

Skim milk powder (SMP) processing is similar to that described above except for the following points:

- contains less milkfat (0.05-0.10%);
- heat treatment prior to evaporation can be more or less severe;
- homogenization not required;
- maximum shelf life extended to approximately three years;

Low-heat SMP is given a pasteurization heat treatment and is used in the production of cheese, baby foods etc. High-heat SMP requires a more intense heat treatment in addition to pasteurization. This product is used in the bakery industry, chocolate industry, and other foods where a high degree of protein denaturation is required.

Whey Powder

Whey is the by-product in the manufacturing of cheese and casein. Disposing of this whey has long been a problem. For environmental reasons it cannot be discharged into lakes and rivers; for economical reasons it is not desirable to simply dump it to waste treatment facilities. Converting whey into powder has led to a number products that it can be incorporated into. It is most desirable, if and where possible, to use it for human food, as it contains a small but valuable protein component. It is also feasible to use it as animal feed. Between the pet food industry and animal feed mixers, hundred's of millions of pounds are sold every year. The feed industry may be the largest consumer of dried whey and whey products.

Whey powder is essentially produced by the same method as other milk powders. Reverse osmosis can be used to partially concentrate the whey prior to vacuum evaporation. Before the whey concentrate is spray dried, lactose crystallization is induced to decrease the hygroscopicity. This is accomplished by quick cooling in flash coolers after evaporation. Crystallization continues in agitated tanks for 4 to 24 h.

A fluidized bed may be used to produce large agglomerated particles with free-flowing, non-hygroscopic, no caking characteristics.

Whey Protein Concentrates

Both whey disposal problems and high-quality animal protein shortages have increased world-wide interest in whey protein concentrates. After clarification and pasteurization, the whey is cooled and held to stabilize the calcium phosphate complex, which later decreases membrane fouling. The whey is commonly processed using ultrafiltration, although reverse osmosis, microfiltration, and demineralization methods can be used. During ultrafiltration, the low molecular weight compounds such as lactose, minerals, vitamins and nonprotein nitrogen are removed in the permeate while the proteins become concentrated in the retentate. After ultrafiltration, the retentate is pasteurized, may be evaporated, then dried. Drying, usually spray drying, is done at lower temperatures than for milk in order that large amounts of protein denaturation may be avoided.

8

Cultured Dairy Products

INTRODUCTION

Traditionally, cheese was made as a way of preserving the nutrients of milk. In a simple definition, cheese is the fresh or ripened product obtained after coagulation and whey separation of milk, cream or partly skimmed milk, buttermilk or a mixture of these products. It is essentially the product of selective concentration of milk. Thousands of varieties of cheeses have evolved that are characteristic of various regions of the world.

Some common cheesemaking steps will be outlined here. Also included is a document entitled Making Cheese at Home, which includes some helpful references, several simple cheese making procedures and information about sourcing cheese making supplies.

- Treatment of Milk
- Additives
- Inoculation and Milk Ripening
- Milk Coagulation
- Enzyme
- Acid

- Heat-acid
- Curd Treatment
- Cheese Ripening

Treatment of Milk for Cheesemaking

Like most dairy products, cheesemilk must first be clarified, separated and standardized. The milk may then be subjected to a sub-pasteurization treatment of 63-65° C for 15 to 16 sec. This thermization treatment results in a reduction of high initial bacteria counts before storage. It must be followed by proper pasteurization. While HTST pasteurization (72° C for 16 sec) is often used, an alternative heat treatment of 60° C for 16 sec may also be used. This less severe heat treatment is thought to result in a better final flavour cheese by preserving some of the natural flora. If used, the cheese must be stored for 60 days prior to sale, which is similar to the regulations for raw milk cheese.

Homogenization is not usually done for most cheesemilk. It disrupts the fat globules and increases the fat surface area where casein particles adsorb. This reults in a soft, weak curd at renneting and increased hydrolytic rancidity.

Additives

The following may all be added to the cheese milk:

- Calcium choride
- Nitrates
- Colour
- Hydrogen peroxide
- Lipases

Calcium choride is added to replace calcium redistributed during pasteurization. Milk coagulation by rennet during cheese making requires an optimum balance among ionic calcium and both soluble insoluble calcium phosphate salts. Because calcium phosphates have reverse solubility with respect to temperature, the heat treatment from pasteurization causes the equilibrium to shift towards insoluble forms and depletes both soluble calcium phosphates and ionic calcium. Near normal equilibrium is restored during 24-48 hours of cold storage, but cheese makers can't wait that long, so $CaCl_2$ is added to restore ionic calcium and improve rennetability. The calcium assists in coagulation and reduces the amount of rennet required.

Sodium or *potassium nitrate* is added to the milk to control the undesirable effects of Clostridium tyrobutyricum in cheeses such as Edam, Gouda, and Swiss.

Because *milk colour* varies from season to season, colour may added to standardize the colour of the cheese throughout the year. Annato, Beta-carotene, and paprika are used.

The addition of *hydrogen peroxide* is sometimes used as an alternative treatment for full pateurization.

Lipases, normally present in raw milk, are inactivated during pasteurization. The addition of kid goat lipases are common to ensure proper flavour development through fat hydrolysis.

Inoculation and Milk Ripening

The basis of cheesemaking relies on the fermentation of lactose by lactic acid bacteria (LAB). The LAB produce lactic acid which lowers the pH and in turn assists coagulation, promotes syneresis, helps prevent spoilage and pathogenic bacteria from growing, contributes to cheese texture, flavour and keeping quality. LAB also produce growth factors which encourages the growth of non-starter organisms, and provides lipases and proteases necessary for flavour development during curing. Further information on LAB and starter cultures can be found in the microbiology section.

After innoculation with the starter culture, the milk is held for 45 to 60 min at 25 to 30° C to ensure the bacteria are active, growing and have developed acidity. This stage is called ripening the milk and is done prior to renneting.

MILK COAGULATION

Coagulation is essentially the formation of a gel by destabilizing the casein micelles causing them to aggregate and form a network which partially immobilizes the water and traps the fat globules in the newly formed matrix. This may be accomplished with:

- enzymes;
- acid treatment;
- heat-acid treatment;

Enzymes

Chymosin, or rennet, is most often used for enzyme coagulation.

Acid Treatment

Lowering the pH of the milk results in casein micelle destabilization or aggregation. Acid curd is more fragile than rennet curd due to the loss of calcium. Acid coagulation can be achieved naturally with the starter culture, or artificially with the addition of gluconodeltalactone. Acid coagulated fresh cheeses may include Cottage cheese, Quark, and Cream cheese.

Heat-Acid Treatment

Heat causes denaturation of the whey proteins. The denatured proteins then interact with the caseins. With the addition of acid, the caseins precipitate with the whey proteins. In rennet coagulation, only 76-78% of the protein is recovered, while in heat-acid coagulation, 90% of protein can be recovered. Examples of cheeses made by this method include Paneer, Ricotta and Queso Blanco.

Curd Treatment

After the milk has gel has been allowed to reach the desired firmness, it is carefully cut into small pieces with knife blades or wires. This shortens the distance and increases the available area for whey to be released. The curd pieces immediately begin to shrink and expel the greenish liquid called whey. This syneresis process is further driven by a cooking stage. The increase in temperature causes the protein matrix to shrink due to increased hydrophobic interactions, and also increases the rate of fermentation of lactose to lactic acid. The increased acidity also contributes to shrinkage of the curd particles. The final moisture content is dependant on the time and temperature of the cook stage. This is important to monitor carefully because the final moisture content of the curd determines the residual amount of fermentable lactose and thus the final pH of the cheese after curing.

When the curds have reached the desired moisture and acidity they are separated from the whey. The whey may be removed from the top or drained by gravity. The curd-whey mixture may also be placed in moulds for draining. Some cheese varieties, such as Colby, Gouda, and Brine Brick include a curd washing which increases the moisture content, reduces the lactose content and final acidity, decreases firmness, and increases openness of texture.

Curd handling from this point on is very specific for each cheese variety. Salting may be achieved through brine as with Gouda, surface salt as with Feta, or vat salt as with cheddar. To acheive the characteritics of cheddar, a cheddaring stage (curd manipulation), milling (cut into shreds), and pressing at high pressure are crucial.

Cheese Ripening

Except for fresh cheese, the curd is ripened, or matured, at various temperatures and times until the characteristic flavour, body and texture profile is achieved. During ripening, degradation of lactose, proteins and fat are carried out by ripening agents. The ripening agents in cheese are:

- bacteria and enzymes of the milk
- lactic culture
- rennet
- lipases
- added moulds or yeasts
- environmental contaminants.

Thus the microbiological content of the curd, the biochemical composition of the curd, as well as temperature and humidity affect the final product. This final stage varies from weeks to years according to the cheese variety.

YOGURT

Yogurt (also spelled yogourt or yoghurt) is a semi-solid fermented milk product which originated centuries ago in Bulgaria. It's popularity has grown and is now consumed in most parts of the world. Although the consistency, flavour and aroma may vary from one region to another, the basic ingredients and manufacturing are essentially consistent:

- Ingredients
- Starter culture
- Manufacturing methods
- Yogurt products and others

Ingredients

Although milk of various animals has been used for yogurt production in various parts of the world, most of the industrialized yogurt production uses cow's milk. Whole milk, partially skimmed milk, skim milk or cream may be used. In order to ensure the development of the yogurt culture the following criteria for the raw milk must be met:

- low bacteria count
- free from antibiotics, sanitizing chemicals, mastitis milk, colostrum, and rancid milk
- no contamination by bacteriophages.

Other yogurt ingredients may include some or all of the following:

- *Other Dairy Products:* concentrated skim milk, nonfat dry milk, whey, lactose. These products are often used to increase the nonfat solids content;
- *Sweeteners:* glucose or sucrose, high-intensity sweeteners (e.g. aspartame);

- *Stabilizers:* gelatin, carboxymethyl cellulose, locust bean Guar, alginates, carrageenans, whey protein concentrate.

Flavours

Fruit Preparations: including natural and artificial flavouring, colour.

Starter Culture

The starter culture for most yogurt production in North America is a symbiotic blend of *Streptococcus salivarius* subsp. thermophilus (ST) and *Lactobacillus delbrueckii* subsp. *bulgaricus* (LB). Although they can grow independantly, the rate of acid production is much higher when used together than either of the two organisms grown individually. The ST grows faster and produces both acid and carbon dioxide. The formate and carbon dioxide produced stimulates LB growth. On the other hand, the proteolytic activity of LB produces stimulatory peptides and amino acids for use by ST. These microorganisms are ultimately responsible for the formation of typical yogurt flavour and texture. The yogurt mixture coagulates during fermentation due to the drop in pH. The streptococci are responsible for the initial pH drop of the yogurt mix to approximately 5.0. The lactobacilli are responsible for a further decrease to pH 4.0. The following fermentation products contibute to flavour:

- lactic acid
- acetaldehyde
- acetic acid
- diacetyl

Manufacturing Method

The milk is clarified and separated into cream and skim milk, then standardized to acheive the desired fat content. The various ingredients are then blended together in a mix tank equipped with a powder funnel and an agitation system. The mixture is then pasteurized using a continuous plate heat exchanger for 30 min at 85° C or 10 min at 95° C. These heat treatments, which are much more severe than fluid milk pasteurization, are necessary to acheive the following:

- produce a relatively sterile and condusive environment for the starter culture;
- denature and coagulate whey proteins to enhance the viscosity and texture.

The mix is then homogenized using high pressures of 2000-2500 psi. Besides thoroughly mixing the stabilizers and other ingredients, homogenization also prevents creaming and wheying off during incubation and storage. Stability, consistency and body are enhanced by homogenization. Once the homogenized mix has cooled to an optimum growth temperature, the yogurt starter culture is added.

A ratio of 1:1, ST to LB, inoculation is added to the jacketed fermentation tank. A temperature of 43° C is maintained for 4-6 h under quiescent (no agitation) conditions. This temperature is a compromise between the optimums for the two micoorganisms (ST 39° C; LB 45° C). The titratable acidity is carefully monitored until the TA is 0.85 to 0.90%. At this time the jacket is replaced with cool water and agitation begins, both of which stop the fermentation. The coagulated product is cooled to 5-22° C, depending on the product. Fruit and flavour may be incorporated at this time, then packaged. The product is now cooled and stored at refrigeration temperatures (5° C) to slow down the physical, chemical and microbiological degradation.

Yogurt Products

There are two types of plain yogurt:

1. Stirred style yogurt; and
2. Set style yogurt.

The above description is essentially the manufacturing proceedures for stirred style. In set style, the yogurt is packaged immediately after inoculation with the starter and is incubated in the packages.

Other yogurt products include:

- Fruit-on-the-bottom style: fruit mixture is layered at the bottom followed by inoculated yogurt, incubation occurs in the sealed cups;
- Soft-serve and Hard Pack frozen yogurt.
- Continental, French, and Swiss: stirred style yogurt with fruit preparation

Yogurt Beverages

Drinking yogurt is essentially stirred yogurt which has a total solids content not exceeding 11% and which has undergone homogenization to further reduce the viscosity. Flavouring and colouring are invariably added. Heat treatment may be applied to extend the storage life. HTST pasteurization with aseptic processing will give a shelf life of several weeks at 2-4°C, which UHT processes with aseptic packaging will give a shelf life of several weeks at room temperature.

OTHER FERMENTED MILK BEVERAGES

Cultured Buttermilk

This product was originally the fermented byproduct of butter manufacture, but today it is more common to produce cultured buttermilks from skim or whole milk. The culture most frequently used in *S. lactis*, perhaps also spp. cremoris. Milk is usually heated to 95°C and cooled to 20-25°C before the addition of the starter culture. Starter is added at 1-2% and the fermentation is allowed to proceed for 16-20 hours, to an acidity of 0.9% lactic acid. This product is frequently used as an ingredient in the baking industry, in addition to being packaged for sale in the retail trade.

Acidophilus Milk

Acidophilus milk is a traditional milk fermented with *Lactobacillus acidophilus* (LA), which has been thought to have therapeutic benefits in the gastrointestinal tract. Skim or whole milk may be used. The milk is heated to high temperature, e.g., 95°C for 1 hour, to reduce the microbial load and favour the slow growing LA culture. Milk is inoculated at a level of 2-5% and incubated at 37°C until coagulated. Some acidophilus milk has an acidity as high as 1% lactic acid, but for therapeutic purposes 0.6-0.7% is more common.

Another variation has been the introduction of a sweet acidophilus milk, one in which the LA culture has been added but there has been no incubation. It is thought that the culture will reach the GI tract where its therapeutic effects will be realized, but the milk has no fermented qualities, thus delivering the benefits without the high acidity and flavour, considered undesirable by some people.

Sour Cream

Cultured cream usually has a fat content between 12-30%, depending on the required properties. The starter is similar to that used for cultured buttermilk. The cream after standardization is usually heated to 75-80°C and is homogenized at >13 MPa to improve the texture. Inoculation and fermentation conditions are also similar to those for cultured buttermilk, but the fermentation is stopped at an acidity of 0.6%.

Others

There are a great many other fermented dairy products, including kefir, koumiss, beverages based on bulgaricus or bifidus strains, labneh, and a host of others. Many of these have developed in regional areas and, depending on the starter organisms used, have various flavours, textures, and components from the fermentation process, such as gas or ethanol.

Whipped Cream Structure

The structure of whipped cream is very similar to the fat and air structure that exists in ice cream. Cream is an emulsion with a fat content of 35-40%. When you whip a bowl of heavy cream, the agitation and the air bubbles that are added cause the fat globules to begin to partially coalesce in chains and clusters and adsorb to and spread around the air bubbles.

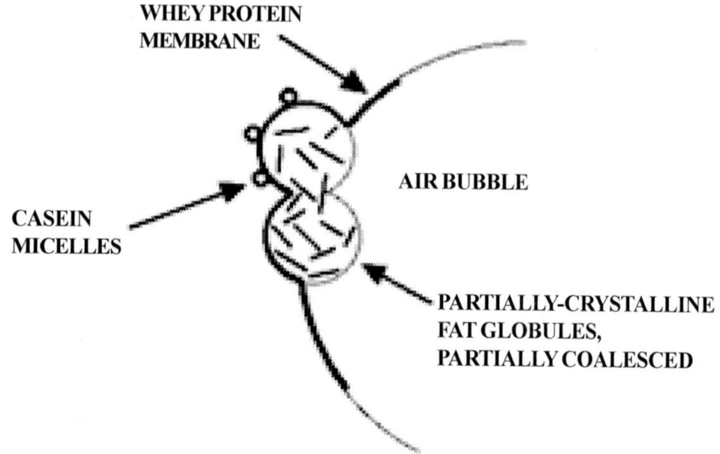

Fig. 8.1: **whipped cream structure**

As the fat partially coalesces, it causes one fat-stabilized air bubble to be linked to the next, and so on. The whipped cream soon starts to become stiff and dry appearing and takes on a smooth texture. This results from the formation of this partially coalesced fat structure stabilizing the air bubbles. The water, lactose and proteins are trapped in the spaces around the fat-stabilized air bubbles. The crystalline fat content is essential (hence whipping of cream is very temperature dependent) so that the fat globules partially coalesce into a 3-dimensional structure rather than fully coalesce into larger and larger globules that are not capable of structure-building. This is caused by the crystals within the globules that cause them to stick together into chains and clusters, but still retain the individual identity of the globules (Fig. 8.2).

Below are scanning electron micrographs image of whipped cream. If you compare the schematics above with the "real thing" below, you should be able to fully understand whipped cream structure.

The structure of whipped cream as determined by scanning electron microscopy.

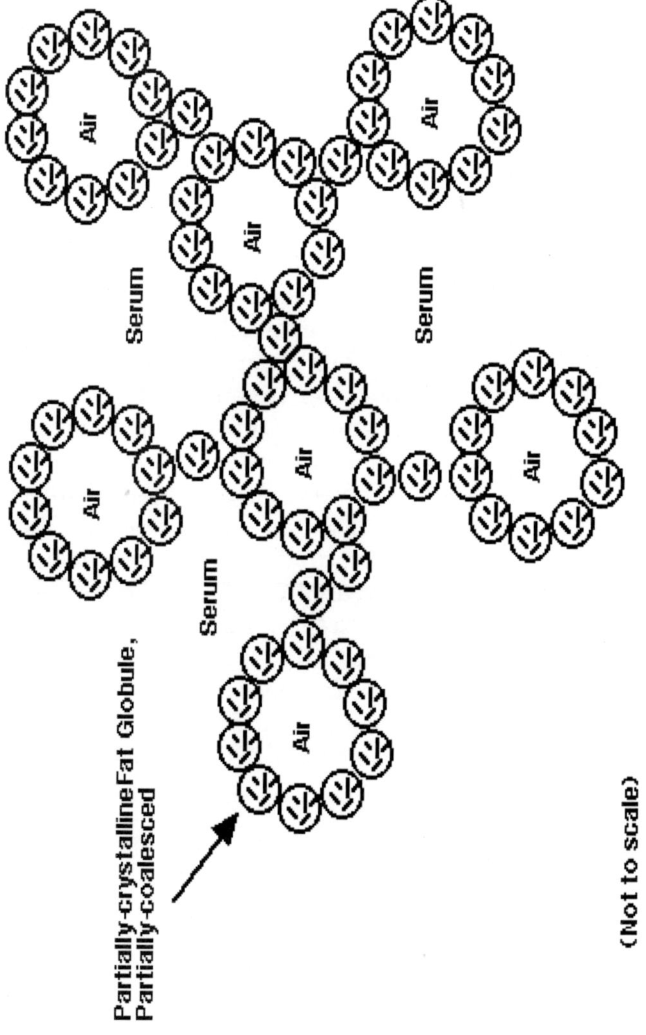
Air
Air
Air
Air
Air
Air
Serum
Serum
Serum
Partially-crystalline Fat Globule,
Partially-coalesced
(Not to scale)

Fig. 8.2

A. Overview showing the relative size and prevalence of air bubbles (a) and fat globules (f); bar = 30 µm.

B. Internal structure of the air bubble, showing the layer of partially coalesced fat which has stabilized the bubble; bar = 5 µm.

C. Details of the partially coalesced fat layer, showing the interaction of the individual fat globules. Bar = 3 µm.

Butter Manufacture

Butter is essentially the fat of the milk. It is usually made from sweet cream and is salted. However, it can also be made from acidulated or bacteriologically soured cream and saltless (sweet) butters are also available. Well into the 19th century butter was still made from cream that had been allowed to stand and sour naturally. The cream was then skimmed from the top of the milk and poured into a wooden tub. Buttermaking was done by hand in butter churns. The natural souring process is, however, a very sensitive one and infection by foreign micro-organisms often spoiled the result. Today's commercial buttermaking is a product of the knowledge and experience gained over the years in such matters as hygiene, bacterial acidifying and heat treatment, as well as the rapid technical development that has led to the advanced machinery now used. The commercial cream separator was introduced at the end of the 19th century, the continuous churn had been commercialized by the middle of the 20th century.

Definitions and Standards

Milkfat

— the lipid components of milk, as produced by the cow, and found in commercial milk and milk-derived products, mostly comprised of triglyceride.

Butterfat

— almost synonymous with milkfat; all of the fat components in milk that are separable by churning.

Anhydrous Milkfat (AMF)

— the commercially-prepared extraction of cow's milkfat, found in bulk or concentrated form (comprised of 100% fat, but not necessarily all of the lipid components of milk).

Butteroil

— synonomous with anhydrous milkfat; (conventional terminology in the fats and oils field differentiates an oil from a fat based on whether it is liquid at room temp. or solid, but very arbitrary).

Butter

— a water-in-oil emulsion, comprised of >80% milkfat, but also containing water in the form of tiny droplets, perhaps some milk solids-not-fat, with or without salt (sweet butter); texture is a result of working/kneading during processing at appropriate temperatures, to establish fat crystalline network that results in desired smoothness (compare butter with melted and recrystallized butter); used as a spread, a cooking fat, or a baking ingredient.

The principal constituents of a normal salted butter are fat (80-82%), water (15.6-17.6%), salt (about 1.2%) as well as protein, calcium and phosphorous (about 1.2%). Butter also contains fat-soluble vitamins A, D and E. Butter should have a uniform colour, be dense and taste clean. The water content should be dispersed in fine droplets so that the butter looks dry. The consistency should be smooth so that the butter is easy to spread and melts readily on the tongue. The buttermaking process involves quite a number of stages. The continuous buttermaker has become the most common type of equipment used.

The cream can be either supplied by a fluid milk dairy or separated from whole milk by the butter manufacturer. The cream should be sweet (pH >6.6, TA = 0.10 - 0.12%), not rancid and not oxidized. If the cream is separated by the butter manufacturer, the whole milk is preheated to the required temperature in a milk pasteurizer before being passed through a separator. The cream is cooled and led to a storage tank where the fat content is analyzed and adjusted to the desired value, if necessary. The skim milk from the separator is pasteurized and cooled before being pumped to storage. It is usually destined for concentration and drying. From the intermediate storage tanks, the cream goes to pasteurization at a temperature of 95°C or more. The high temperature is needed to destroy enzymes and micro-organisms that would impair the keeping quality of the butter.

If ripening is desired for the production of cultured butter, mixed cultures of S. cremoris, S. lactis diacetyl lactis, Leuconostocs, are used and the cream is ripened to pH 5.5 at 21°C and then pH 4.6 at 13°C. Most flavour development occurs between pH 5.5-4.6. The colder the temperature during ripening the more the flavour development relative to acid production. Ripened butter is usually not washed or salted. In the aging tank, the cream is subjected to a program of controlled cooling designed to give the fat the required crystalline structure. The program is chosen to accord with factors such as the composition of the butterfat, expressed, for example, in terms of the iodine value which is a measure of the unsaturated fat content. The treatment can even be modified to obtain butter with good consistency despite a low iodine value, i.e. when the unsaturated proportion of the fat is low. As a rule, aging takes 12-15 hours. From the aging tank, the cream is pumped to the churn or continuous buttermaker via a plate heat exchanger which brings it to the requisite temperature. In the churning process the cream is violently agitated to break down the fat globules, causing the fat to coagulate into butter grains, while the fat content of the remaining liquid, the buttermilk, decreases.

Thus the cream is split into two fractions: butter grains and buttermilk. In traditional churning, the machine stops when the grains have reached a certain size, whereupon the buttermilk is drained off. With the continuous buttermaker the draining of the buttermilk is also continuous. After draining, the butter is worked to a continuous fat phase containing a finely dispersed water phase. It used to be common practice to wash the butter after churning to remove any residual buttermilk and milk solids but this is rarely done today. Salt is used to improve the flavour and the shelf-life, as it acts as a preservative. If the butter is to be salted, salt (1-3%) is spread over its surface, in the case of batch production. In the continuous buttermaker, a salt slurry is added to the butter. The salt is all dissolved in the aqueous phase, so the effective salt concentration is approximately 10% in the water.

After salting, the butter must be worked vigorously to ensure even distribution of the salt. The working of the butter also influences the characteristics by which the product is judged - aroma, taste, keeping quality, appearance and colour. Working is required to obtain a homogenous blend of butter granules, water and salt. During working, fat moves from globular to free fat. Water droplets decrease in size during working and should not be visible in properly worked butter. Overworked butter will be too brittle or greasy depending on whether the fat is hard or soft. Some water may be added to standardize the moisture content. Precise control of composition is essential for maximum yield. The finished butter is discharged into the packaging unit, and from there to cold storage.

THE BACKGROUND SCIENCE OF BUTTER CHURNING

The Fat Globule

Milk fat is comprised mostly of triglycerides, with small amounts of mono- and diglycerides, phospholipids, glycolipids, and lipo-proteins. The trigylcerides (98% of milkfat) are of diverse composition with respect to their component fatty acids, approximately 40% of which are unsaturated fat firmness varies with chain length, degree of unsaturation, and position of the fatty acids on the glycerol. Fat globules vary from 0.1-10 micron in diameter. The fat globule membrane is comprised of surface active materials: phospholipids and lipoproteins.

Fat globules typically aggregate in three ways:

- flocculation;
- coalescence; and
- partial coalescence;

Whipping and Churning

Many milk products foam easily. Skim milk foams copiously with the amount of foam being very dependent on the amount of residual fat - fat depresses foaming. The foaming agents are proteins, the amount of proteins in the foam are proportional to their contents in milk. Foaming is decreased in heat treated milk, possibly because denaturated whey proteins produce a more brittle protein layer at the interface. Fats tend to spread over the air-water interface and destabilize the foam; very small amounts of fats (including phospholipids) can destabilize a foam.

During the interaction of fat globules with air bubbles the globule may also be disrupted (this is the only way that fat globules can be disrupted without considerable energy input). Disruption of the fat globule by interaction between the fat globule and air bubbles is rare except in the case of newly formed air bubbles where the air-water interfacial layer is still thin. If part of the fat globule is solid, churning will result, hence the term "flotation churning" -from repeated rupturing of air bubbles and resulting coalescence of the adsorbed fat. In spite of the above comments on the destabilization of foams by fat, milk fat is essential for the formation of stable whipped products which depend on the interaction between fat globules, air bubbles and plasma components (esp. proteins).

When cream is beaten air cells form more slowly partly because of higher viscosity and partly because the presence of fat causes immediate collapse of most of the larger bubbles. If most of the fat is liquid (high temperature) the fat globule membrane is not readily punctured and churning does not occur -at cold temperature where solid fat is present, churning (clumping) of the fat globule takes place. Clumps of globules begin to associate with air bubbles so that a network of air bubbles and fat clumps and globules form entrapping all the liquid and producing a stable foam. If beating continues the fat clumps increase in size until they become too large and too few to enclose the air cells, hence air bubbles coalesce, the foam begins to "leak" and ultimately butter and butter milk remain.

Crystallizing of the milkfat during Aging

Before churning, cream is subjected to a programme of cooling designed to control the crystallization of the fat so that the resultant butter has the right consistency. The consistency of butter is one of its most important quality-related characteristics, both directly and indirectly, since it affects the other characteristics - chiefly taste and aroma. Consistency is a complicated concept and involves properties such as hardness, viscosity, plasticity and spreading ability.

The relative amounts of fatty acids with high melting point determine whether the fat will be hard or soft. Soft fat has a high content of low-melting fatty acids and at room temperature this fat has a large continuous fat phase with a low solid phase, i.e. crystallized, high-melting fat. On the other hand, in a hard fat, the solid phase of high-melting fat is much larger than the continuous fat phase of low-melting fatty acids.

In buttermaking, if the cream is always subjected to the same heat treatment it will be the chemical composition of the milk fat that determines the butter's consistency. A soft milk fat will make a soft and greasy butter, whereas butter from hard milk fat will be hard and stiff. If, however, the heat treatment is modified to suit the iodine value of the fat, the consistency of the butter can be optimized. For the heat treatment regulates the size of the fat crystals, and the relative amounts of solid fat and the continuous phase - the factors that determine the consistency of the butter.

Pasteurization causes the fat in the fat globules to liquefy. And when the cream is subsequently cooled a proportion of the fat will crystallize. If cooling is rapid, the crystals will be many and small; if gradual the yield will be fewer but larger crystals. The more violent the cooling process, the more will be the fat that will crystallize to form the solid phase, and the less the liquid fat that can be squeezed out of the fat globules during churning and working.

The crystals bind the liquid fat to their surface by adsorption. Since the total surface area is much greater if the crystals are many and small, more liquid fat will be adsorbed than if the crystals were larger and fewer. In the former case, churning and working will press only a small proportion of the liquid fat from the fat globules. The continuous fat phase will consequently be small and the butter firm. In the latter case, the opposite applies. A larger amount of liquid fat will be pressed out; the continuous phase will be large and the butter soft. So by modifying the cooling program for the cream, it is possible to regulate the size of the crystals in the fat globules and in this way influence both the magnitude and the nature of the important continuous fat phase.

Treatment of Hard Fat

For optimum consistency where the iodine value is low, i.e. the butterfat is hard, as much as possible of the hardest fat must be converted to as few crystals as possible, so that little of the liquid fat is bound to the crystals. The liquid fat phase in the fat globules will thereby be maximized and much of it can be pressed out during churning and working, resulting in butter with a relatively large continuous phase of liquid fat and with the hard fat concentrated to the solid phase. The programme of treatment necessary to achieve this result comprises the following stages:

— rapid cooling to about 8°C and storage for about 2 hours at this temperature;

— heating gently to 20-21°C and storage at this temperature for at least 2 hours (water at 27-29°C is used for heating);

— cooling to about 16°C.

Cooling to about 8°C causes the formation of a large number of small crystals that bind fat from the liquid continuous phase to their surface. When the cream is gently heated to 20-21°C the bulk of the crystals melt, leaving only the hard fat crystals which, during the storage period at 20-21°C, grow larger.

After 1-2 hours most of the hard fat has crystallized, binding little of the liquid fat. By dropping the temperature now to about 16°C, the hardest portion of the fat will be fixed in crystal form while the rest is liquefied. During the holding period at 16°C, fat with a melting point of 16°C or higher will be added to the crystals. The treatment has thus caused the high-melting fat to collect in large crystals with little adsorption of the low-melting liquid fat, so that a large proportion of the butter oil can be pressed out during churning and working.

Treatment of Medium-hard Fat

With an increase in the iodine value, the heating temperature is accordingly reduced from 20-21°C. Consequently a larger number of fat crystals will form and more liquid fat will be adsorbed than is the case with the hard fat program. For iodine values up to 39, the heating temperature can be as low as 15°C.

Treatment of Very Soft Fat

Where the iodine value is greater than 39-40 the "summer method" of treatment is used. After pasteurization the cream is cooled to 20°C. If the iodine value is around 39-40 the cream is cooled to about 8°C, and if 41 or greater to 6°C. It is generally held that aging temperatures below the 20° level will give a soft butter.

Continuous Buttermaking

There are essentially four types of buttermaking processes:

1. traditional batch churning from 25-35% mf. cream;
2. continuous flotation churning from 30-50% mf. cream;
3. the concentration process whereby "plastic" cream at 82% mf. is separated from 35% mf. cream at 55ºC and then

this oil-in-water emulsion cream is inverted to a water-in-oil emulsion butter with no further draining of buttermilk; and

4. the anhydrous milkfat process whereby water, SNF, and salt are emulsified into butter oil in a process very similar to margarine manufacture.

An optimum churning temperature must be determined for each type of process but is mainly dependent on the mean melting point and melting range of the lipids, as discussed above, i.e., 7-10°C in summer and 10-13°C in winter. If churning temperature is too warm or if the thermal cream aging cycle permits too much liquid fat, then a soft greasy texture results; if too cold or too much solid fat, then butter becomes too brittle. The cream is first fed into a churning cylinder fitted with beaters that are driven by a variable speed motor.

Rapid conversion takes place in the cylinder and, when finished, the butter grains and buttermilk pass on to a draining section. The first washing of the butter grains sometimes takes place en route-either with water or recirculated chilled buttermilk. The working of the butter commences in the draining section by means of a screw, which also conveys it to the next stage.

On leaving the working section the butter passes through a conical channel to remove any remaining buttermilk. Immediately afterwards, the butter may be given its second washing, this time by two rows of adjustable high-pressure nozzles. The water pressure is so high that the ribbon of butter is broken down into grains and consequently any residual milk solids are effectively removed. Following this stage, salt may be added through a high-pressure injector. The third section in the working cylinder is connected to a vacuum pump. Here it is possible to reduce the air content of the butter to the same level as conventionally churned butter. In the final or mixing section the butter passes a series of perforated disks and star wheels. There is also an injector for final adjustment of the water content. Once regulated, the water content of the butter deviates less than + 0.1%, provided the characteristics

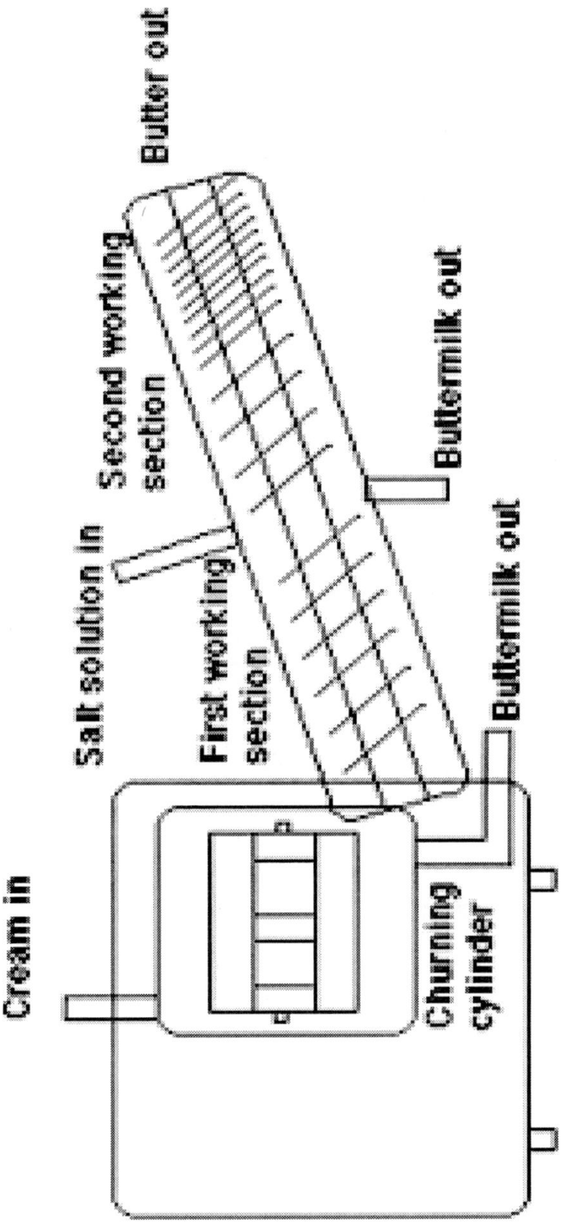
Cream in
Churning cylinder
Salt solution in
First working section
Second working section
Butter out
Buttermilk out
Buttermilk out

Fig. 8.3

of the cream remain the same. The finished butter is discharged in a continuous ribbon from the end nozzle of the machine and then into the packaging unit.

Concentration Method

- 30% fat cream pasteurized at 90ºC
- degassed in a vacuum
- cooled to 45-70°C
- separated to 82% fat ("plastic" cream)
- the concentrate, still an O/W emulsion, is cooled to 8-13°C
- fat crystals forming in the tightly packed globules perforate the membranes, cause liquid fat leakage and rapid phase inversion
- contrast to mayonnaise, also a O/W emulsion at 82% fat but is winterized to prevent crystallization
- butter from this method contains all membrane material, therefore, more phospholipids
- no butter milk produced
- after phase inversion the butter is worked and salted.

Phase Separation

Butter from anhydrous milkfat:

- prepare "plastic" cream (>80% fat)
- heat with agitation to destabilize emulsion
- separate oil from aqueous phase: 82 to 98% butter fat
- this butter oil is then blended with water, salt and milk solids in an emulsion pump and transferred to a scraped surface heat exchanger for cooling and to initiate crystallization
- further worked to develop crystal structure and texture
- process similar to margarine manufacture

- margarine has advantage of fat composition control to modify physical properties
- butter produced by phase separation contains few phospholipids.

Butter Yield Calculations

Technological limits to yield efficiency are defined by separation efficiency, churning efficiency, composition overrun, and package over fill.

1. Separation efficiency (Es)

— represents fat transferred from milk to cream

Es = 1 - fs/fm

where fs = skim fat as per cent

w/w fm = milk fat as percent w/w

Separation efficiency depends on initial milk fat content and residual fat in the skim. Assuming optimum operation of the separator, the principal determining factor of fat loss to the skim is fat globule size. Modern separators should achieve a skim fat content of 0.04 - 0.07%.

2. Churning Efficiency (Ec):

— represents fat transferred from cream to butter

Ec = 1 - fbm/fc

where fbm = buttermilk fat as percent w/w

fc = cream fat as percent w/w

Maximum acceptable fat loss in buttermilk is about 0.7% of churned fat corresponding to a churning efficiency of 99.3% of cream fat recovered in the butter. Churning efficiency is highest in the winter months and lowest in the summer months. Fat losses are higher in ripened butter due to a restructuring of the FGM (possibly involving crystallization of high melting triglycerides on the surface of the globules). If churning temperature is too high, churning occurs more quickly but fat loss in buttermilk increases. For continuous churns assuming 45% cream, churning efficiency should be 99.61 - 99.42%.

3. Composition Overrun

% Churn Overrun

= (Kg butter made - Kg fat churned)/Kg fat churned x 100 %

% Composition Overrun

= (100 - % fat in butter)/% fat × 100 %

4. Package Fill Control

= (actual wt. - nominal wt.)/nominal wt. x 100%

An acceptable range for 25 kg butter blocks is 0.2 - 0.4% ovefill. Overfill on 454 g prints is about 0.6%.

5. Other factors affecting yield

— shrinkage due to leaky butter (improperly worked).

— shrinkage due to moisture loss; avoided by aluminum wrap.

— loss of butter remnants on processing equipment; % loss minimal in large scale continuous processing.

6. Plant Overrun

— Plant efficiency or plant overrun is the sum of separation, churning, composition overrun and package fill efficiencies. In summary the theoretical maximum efficiency values are:

Separation Efficiency 98.85

Churning Efficiency 99.60

Composition overrun (% fat) 23.30

Package overfill 0.20

— These values can be used to predict the expected yield of butter per kg of milk or kg of milk fat received.

7. Example

3.6% m.f. milk

0.05% m.f. in skim

40% m.f. in cream

0.3% m.f. in buttermilk

81.5% m.f. in butter

Es = 1 - .05/3.6 = 98.6

Ec = 1 - .3/40 = 99.25

% Composition Overrun = (100-81.5)/81.5 = 22.7%

If 100 kg of milk was used, 8.9 kg of cream would be produced (from a Pearson Square mass balance) and 4.35 kg butter would be produced from that. This is the theoretical yield based on no losses. The mass balance of fat shows that 98.3% of the fat ended up in the butter, 0.4% of the fat ended up in the buttermilk and 1.3% of the fat ended up in the skim.

The % Churn Overrun = (4.35 - 3.6)/3.6 = 20.8%

WHIPPED BUTTER

Whipped butter is typically used in foodservice situations. The main advantage of whipped butter is increased spreadability even at refrigeration temperatures, thus providing great advantage for the restaurant industry. The volume increase is usually 25-30%. Whipping is achieved by injecting an inert gas (nitrogen) into the butter after churning. In the phase separation process, whipping can be achieved by injecting nitrogen in the crystallizer as is done in the production of whipped margarine.

Anhydrous Milkfat ("butter oil")

Anhydrous milk fat, butter oil, can be manufactured from either butter or from cream. For the manufacture from butter, non-salted butter from sweet cream is normally used, and the process works better if the butter is at least a few weeks old. Melted butter is passed through a centrifuge, to concentrate the fat to 99.5% of greater. This oil is heated again to 90-95°C and vacuum cooled before packaging.

The processes for the production of anhydrous fat, using cream as the raw material, are based on the emulsion splitting principle. In brief, the processes consist of the cream first being concentrated to 75% fat or greater, in two stages. In both of these stages, the fat is concentrated in a hermetic solids-ejecting separator. The fat globules are then broken down mechanically, so that phase inversion occurs and the fat is liberated. This forms a continuous fat phase containing dispersed water droplets, which can be separated from the fat phase by centrifugation. This is similar to the concentration method for buttermaking, with the addition of the mechanical rupture of the emulsion and additional separator for removal of the residual water phase.

One of the key machines in the system is the mechanical device for phase inversion. This can be in the form of a centrifugal separator equipped with a serrated disc. The disc breaks down the emulsion, so that the liquid leaving the machine is a continuous oil phase, with dispersed water droplets and buttermilk. Larger equipment could be equipped with a motor-driven serrated disc or with a homogenizer. After phase inversion, the fat is concentrated to 99.5% or greater in a hermetic separator.

FRACTIONATION OF ANHYDROUS MILK FAT

Milk fat is a complicated mixture of triglycerides that contain numerous fatty acids of varying carbon chain lengths and degrees of saturation. The proportions of the various fatty acids present will also vary depending on the conditions surrounding the production of milk. One method of milkfat fraction is by thermal treatment. The mixture can be separated into fractions on the basis of their melting point.

The technique consists of melting the entire quantity of fat and then cooling it down to a predetermined temperature. The triglycerides with the higher melting point will then crystallize and settle out. In the modern thermal fractionation method, sedimentation by gravity is replaced by centrifugal separation. Since a modern separator generates a force that is thousands of times greater than the force of gravity and since the sedimentation distances are very short, the process is incomparably faster.

The crystallizing stage can also be accelerated, since the crystals need not be large if centrifugal separation is employed. Fractionation of milkfat can also be accomplished by supercritical fluid extraction techniques.

9

MILK CONTENTS

INTRODUCTION

Milk is an opaque white liquid produced by the mammary glands of female mammals (including monotremes). It provides the primary source of nutrition for newborn mammals before they are able to digest other types of food. The early lactation milk is known as colostrum, and carries the mother's antibodies to the baby. It can reduce the risk of many diseases in the baby. The exact components of raw milk varies by species, but it contains significant amounts of saturated fat, protein and calcium as well as vitamin C. Cow's milk has a pH ranging from 6.4 to 6.8, making it slightly acidic.

Types of Consumption

There are two distinct types of milk consumption: a natural source of nutrition for all infant mammals, and a food product for humans of all ages derived from other animals.

Nutrition for Infant Mammals

A goat kid feeding on its mother's milk.In almost all mammals, milk is fed to infants through breastfeeding, either directly or, for humans, by expressing the milk to be stored and consumed later. Some cultures, historically or currently, continue to use breast milk to feed their children until they are 7-year old.

Food Product for Humans

In many cultures of the world, especially the Western world, humans continue to consume milk beyond infancy, using the milk of other animals (in particular, cows) as a food product. For millennia, cow milk has been processed into dairy products such as cream, butter, yogurt, kefir, ice cream, and especially the more durable and easily transportable product, cheese. Industrial science has brought us casein, whey protein, lactose, condensed milk, powdered milk, and many other food-additive and industrial products.

Humans are an exception in the natural world for consuming milk past infancy. Most humans lose the ability to fully digest milk after childhood (that is, they become lactose intolerant). The sugar lactose is found only in milk, forsythia flowers, and a few tropical shrubs. The enzyme needed to digest lactose, lactase, reaches its highest levels in the small intestines after birth and then begins a slow decline unless milk is consumed regularly. On the other hand, those groups that do continue to tolerate milk often have exercised great creativity in using the milk of domesticated ungulates, not only of cows, but also sheep, goats, yaks, water buffalo, horses, and camels.

Milking has its advent in the very evolution of placental mammals. While the exact time of its appearance is not known, the immediate ancestors of modern mammals were much like monotremes, including the platypus. Such animals today produce a milk-like substance from glands on the surface of their skin, but without the nipple, for their offspring to drink after hatching from their eggs. Likewise, marsupials, the closest cousin to placental mammals, produce a milk-like substance from a teat-like organ in their pouches. The earliest immediate ancestor of placental mammals known seems to be eomaia, a small creature superficially resembling rodents, that is thought to have lived 125 million years ago, during the Cretaceous era. It almost certainly produced what would be considered milk, in the same way as modern placental mammals.

Animal milk is first known to have been used as human food at the beginning of animal domestication. Cow milk was first used as human food in the Middle East. Goats and sheep were domesticated in the Middle East between 9000 and 8000 BC Goats and sheep are ruminants: mammals adapted to survive on a diet of dry grass, a food source otherwise useless to humans, and one that is easily stockpiled. The animals were probably first kept for meat and hides, but dairying proved to be a more efficient way of turning uncultivated grasslands into sustenance: the food value of an animal killed for meat can be matched by perhaps one year's worth of milk from the same animal, which will keep producing milk — in convenient daily portions — for years

Around 7000 BC, cattle were being herded in parts of Turkey. There is evidence from DNA extraction of skeletons from the Neolithic period that people in northern Europe were missing the necessary genes to process lactase. Scientists claim it is more likely that the genetic mutation allowing the digestion of milk arose at some point after dairy farming began The use of cheese and butter spread in Europe, parts of Asia and parts of Africa. Domestic cows, which previously existed throughout much of Eurasia, were then introduced to the colonies of Europe during the Age of exploration.

Milk was first delivered in bottles on January 11, 1878. The day is now remembered as Milk Day and is celebrated annually. The town of Harvard, Illinois also celebrates milk in the summer with a festival known as Milk Days. Theirs is a different tradition meant to celebrate dairy farmers in the "Milk Capital of the World."

In addition to cows, the following animals provide milk used by humans for dairy products:

- Buffalo
- Camels
- Donkeys
- Goats (the nanny)
- Horses (the mare)

- Reindeer
- Sheep (the ewe)
- Water buffalo
- Yaks

In Russia and Sweden, small moose dairies also exist. Donkey and horse milk have the lowest fat content, while the milk of seals contains more than 50% fat.

Whale milk, not used for human consumption, is one of the highest-fat milks, containing up to 50% fat. The high fat content of whale milk is not unique to the cetacean's great size, as guinea pig milk has an average fat content of 46%.

Human milk is not produced or distributed industrially or commercially; however, milk banks exist that allow for the collection of donated human milk and its redistribution to infants who may benefit from human milk for various reasons (premature neonates, babies with allergies or metabolic diseases, etc.)

All other female mammals do produce milk, but are rarely or never used to produce dairy products for human consumption.

Modern Production

In the Western world today, cow milk is produced on an industrial scale. It is by far the most commonly consumed form of milk in the western world. Commercial dairy farming using automated milking equipment produces the vast majority of milk in developed countries. Types of cattle such as the Holstein have been specially bred for increased milk production. 90% of the dairy cows in the United States and 85% in Great Britain are Holsteins. Other milk cows in the United States include Ayrshire, Brown Swiss, Guernsey, Jersey, and Milking Shorthorn. The largest producers of dairy products and milk today are India followed by the United States and China. In India, Amul, a cooperative owned jointly by 2.6 million small farmers was the engine behind the success of Operation Flood.

Top ten buffalo milk producers — 11 June 2008

Country	*Production (1,000 tonnes)*	*Footnote*
India	56,960	*
Pakistan	21,500	P
People's Republic of China	2,900	F
Egypt	2,300	F
Nepal	930	F
Iran	241.5	F
Myanmar	205	F
Italy	200	F
Turkey	35.1	F
Vietnam	31	F
World	**85396902**	**A**

No symbol = official figure, P = official figure, F = FAO estimate, * = Unofficial/Semi-official/mirror data, C = Calculated figure A = Aggregate (may include official, semi-official or estimates);

Source: Food and Agricultural Organization of United Nations: Economic And Social Department: The Statistical Division.

Price

It was reported in 2007 that with increased worldwide prosperity and the competition of biofuel production for feedstocks, both the demand for and the price of milk had substantially increased world wide. Particularly notable was the rapid increase of consumption of milk in China and the rise of the price of milk in the Uited States above the government subsidized price.

Physical and Chemical Structure

Milk is an emulsion or colloid of butterfat globules within a water-based fluid. Each fat globule is surrounded by a membrane consisting of phospholipids and proteins; these emulsifiers keep the individual globules from joining together into noticeable grains of butterfat and also protect the globules from the fat-digesting activity of enzymes found in the fluid portion of the milk. In unhomogenized cow milk, the fat globules average about four micrometers across. The fat-soluble vitamins A, D, E, and K are found within the milkfat portion of the milk. The largest structures in the fluid portion of the milk are casein protein micelles: aggregates of several thousand protein molecules, bonded with the help of nanometer-scale particles of calcium phosphate. Each micelle is roughly spherical and about a tenth of a micrometer across. There are four different types of casein proteins, and collectively they make up around 80 percent of the protein in milk, by weight. Most of the casein proteins are bound into the micelles. There are several competing theories regarding the precise structure of the micelles, but they share one important feature: the outermost layer consists of strands of one type of protein, kappa-casein, reaching out from the body of the micelle into the surrounding fluid. These Kappa-casein molecules all have a negative electrical charge and therefore repel each other, keeping the micelles separated under normal conditions and in a stable colloidal suspension in the water-based surrounding fluid.

Both the fat globules and the smaller casein micelles, which are just large enough to deflect light, contribute to the opaque white color of milk. Skimmed milk, however, appears slightly blue because casein micelles scatter the shorter wavelengths (blue compared to red).

The fat globules contain some yellow-orange carotene, enough in some breeds — Guernsey and Jersey cows, for instance — to impart a golden or "creamy" hue to a glass of milk. The riboflavin in the whey portion of milk has a greenish color, which can sometimes be discerned in skim milk or whey products. Fat-free skim milk has only the casein micelles to scatter light, and they tend to scatter shorter-wavelength blue light more than they do red, giving skim milk a bluish tint.

Milk contains dozens of other types of proteins besides the caseins. They are more water-soluble than the caseins and do not form larger structures. Because these proteins remain suspended in the whey left behind when the caseins coagulate into curds, they are collectively known as whey proteins. Whey proteins make up around twenty percent of the protein in milk, by weight. Lactoglobulin is the most common whey protein by a large margin.

The carbohydrate lactose gives milk its sweet taste and contributes about 40% of whole cow milk's calories. Lactose is a composite of two simple sugars, glucose and galactose. In nature, lactose is found only in milk and a small number of plants Other components found in raw cow milk are living white blood cells. Mammary-gland cells, various bacteria, and a large number of active enzymes are some other components in milk.

Processing

In most Western countries, a centralised dairy facility processes milk and products obtained from milk (dairy products), such as cream, butter, and cheese. In the United States, these dairies are usually local companies, while in the southern hemisphere facilities may be run by very large nationwide or trans-national corporations (such as Fonterra).

Pasteurization

Pasteurization is used to kill harmful microorganisms by heating the milk for a short time and then cooling it for storage and transportation. Pasteurized milk is still perishable and must be stored cold by both suppliers and consumers. Dairies print expiration dates on each container, after which stores will remove any unsold milk from their shelves. In many countries it is illegal to sell milk that is not pasteurized.

Milk may also be further heated to extend its shelf life through ultra-high temperature treatment (UHT), which allows it to be stored unrefrigerated, or an even longer lasting sterilization process.

Creaming and Homogenization

Upon standing for 12 to 24 hours, fresh milk has a tendency to separate into a high-fat cream layer on top of a larger, low-fat milk layer. The cream is often sold as a separate product with its own uses; today the separation of the cream from the milk is usually accomplished rapidly in centrifugal cream separators. The fat globules rise to the top of a container of milk because fat is less dense than water. The smaller the globules, the more other molecular-level forces prevent this from happening. In fact, the cream rises in cow milk much more quickly than a simple model would predict: rather than isolated globules, the fat in the milk tends to form into clusters containing about a million globules, held together by a number of minor whey proteins These clusters rise faster than individual globules can. The fat globules in milk from goats, sheep, and water buffalo do not form clusters so readily and are smaller to begin with; cream is very slow to separate from these milks.

Milk is often homogenized, a treatment which prevents a cream layer from separating out of the milk. The milk is pumped at high pressures through very narrow tubes, breaking up the fat globules through turbulence and cavitation A greater number of smaller particles possess more total surface area than a smaller number of larger ones, and the original fat globule membranes cannot completely cover them. Casein micelles are attracted to the newly-exposed fat surfaces; nearly one-third of the micelles in the milk end up participating in this new membrane structure. The casein weighs down the globules and interferes with the clustering that accelerated separation. The exposed fat globules are briefly vulnerable to certain enzymes present in milk, which could break down the fats and produce rancid flavors. To prevent this, the enzymes are inactivated by pasteurizing the milk immediately before or during homogenization.

Homogenized milk tastes blander but feels creamier in the mouth than unhomogenized; it is whiter and more resistant to developing off flavors Creamline, or cream-top, milk is unhomogenized; it may or may not have been pasteurized. Milk which has undergone high-pressure homogenization, sometimes labeled as "ultra-homogenized," has a longer shelf life than milk which has undergone ordinary homogenization at lower pressures Homogenized milk may be more digestible than unhomogenized milk.

Concerns exist about the health effects of consuming homogenized milk. Work by Kurt A. Oster, M.D. in the 1960s through the 1980s suggested a link between homogenized milk and arterosclerosis, due to the release of bovine xanthine oxidase (BXO) from the milk fat globular membrane (MFGM) during homogenization. While Oster's work has been widely criticized, it is apparent that homogenization introduces changes the MFGM and the exposure of its proteins, and the effects of these changes on food safety have not been thoroughly investigated.

Nutrition and Health

The composition of milk differs widely between species. Factors such as the type of protein; the proportion of protein, fat, and sugar; the levels of various vitamins and minerals; and the size of the butterfat globules and the strength of the curd are among those than can vary. For example:

- Human milk contains, on average, 1.1% protein, 4.2% fat, 7.0% lactose (a sugar), and supplies 72 kcal of energy per 100 grams.
- Cow milk contains, on average, 3.4% protein, 3.6% fat, and 4.6% lactose, 0.7% minerals and supplies 66 kcal of energy per 100 grams. See also Nutritional value further on.

Aquatic mammals, such as seals and whales, produce milk that is very rich in fats and other solid nutrients when compared with land mammals' milk.

Processed milk began containing differing amounts of fat during the 1950s. A serving (1 cup or 250 ml) of 2%-fat milk contains 285 mg of calcium, which represents 22% to 29% of the daily recommended intake (DRI) of calcium for an adult. Depending on the age, 8 grams of protein, and a number of other nutrients (either naturally or through fortification):

- Biotin and pantothenic acid are B vitamins important for energy production.
- Iodine is a mineral essential for thyroid function.

- Potassium and magnesium are for cardiovascular health.
- Selenium is a cancer-preventive trace mineral.
- Thiamine is a B-vitamin important for cognitive function, especially memory.
- Vitamin A is critical for immune function.
- Vitamin B12 and riboflavin are necessary for cardiovascular health and energy production.
- Vitamins D and K are essential for bone health.

The amount of calcium from milk that is absorbed by the human body is disputed Calcium from dairy products has a greater bioavailability than calcium from certain vegetables, such as spinach, that contain high levels of calcium-chelating agents, but a similar or lesser bioavailability than calcium from low-oxalate vegetables such as kale, broccoli, or other vegetables in the Brassica genus.

Infants fed solely goat milk-based formula with no folic acid supplementation can suffer from folate deficiency, as goat milk has only one-tenth as much folic acid as cow's milk.

Studies show possible links between low-fat milk consumption and reduced risk of arterial hypertension, coronary heart disease, colorectal cancer and obesity. Overweight individuals who drink milk may benefit from decreased risk of insulin resistance and type 2 diabetes One study has shown that for women desiring to have a child, those who consume full fat dairy products may actually slightly increase their fertility, while those consuming low fat dairy products may slightly reduce their fertility due to interference with ovulation. However, studies in this area are still inconsistent. Milk is a source of Conjugated linoleic acid, a fatty acid that inhibits several types of cancer in mice. CLA has been shown to kill human skin cancer, colorectal cancer and breast cancer cells in vitro studies, and may help lower cholesterol and prevent atherosclerosis; CLA is present only in milk from grass-fed cows.

Other studies suggest that milk consumption may increase the risk of suffering from certain health problems. Milk contains casein, a substance that breaks down in the human stomach to produce casomorphin, an opioid peptide. In the early 1990s it was hypothesized that casomorphin can cause or aggravate autism and casein-free diets are widely promoted. Studies supporting these claims have had significant flaws, and the data are inadequate to guide autism treatment recommendations Cow milk allergy (CMA) is as an immunologically mediated adverse reaction to one or more cow milk proteins. Rarely is it severe enough to cause death Studies described in the book The China Study note a correlation between casein intake and the promotion of cancer cell growth when exposed to carcinogens. However other studies have shown whey protein offers a protective effect against colon cancer.

A study demonstrated that men, and to some degree women, who drink a large amount of milk and consume dairy products were at a slightly increased risk of developing Parkinson's disease. The reason behind this is not fully understood, and it also remains unclear why there is less of a risk for women. Several sources suggest a correlation between high calcium intake (2000 mg per day, or twice the US recommended daily allowance, equivalent to six or more glasses of milk per day) and prostate cancer A large study specifically implicates dairy. A review published by the World Cancer Research Fund and the American Institute for Cancer Research states that at least eleven human population studies have linked excessive dairy product consumption and prostate cancer, however randomized clinical trial data with appropriate controls only exists for calcium, not dairy produce, where there was no correlation. Medical studies have also shown a possible link between milk consumption and the exacerbation of diseases such as Crohn's Disease, Hirschsprung's disease–mimicking symptoms in babies with existing cow milk allergies severe gastroesophageal reflux disease in infants and children hypersenstitive to milk, and the aggravation of Behçet's disease.

Since November 1993, with FDA approval, Monsanto has been selling recombinant bovine somatotropin (rbST)—or rBGH—to dairy farmers. Additional bovine growth hormone is administered to cattle in order to increase their milk production, though the hormone also naturally fosters liver production of insulin-like growth factor 1 (IGF1). The deposit thereof in the milk of rBGH-affected cattle has been the source of concern; however, all milk contains IGF1 since all milking cows produce bovine growth hormone naturally. The IGF1 in milk from rBGH-affected cattle does not vary from the range normally found in a non-supplemented cow Elevated levels of IGF1 in human blood has been linked to increased rates of breast, colon, and prostate cancer by stimulating their growth, though this has not been linked to milk consumption. The EU has recommended against Monsanto milk n addition, the cows receiving rBGH supplements may more frequently contract an udder infection known as mastitis. Milk from rBGH-affected cattle is banned in Canada, Australia, New Zealand, and Japan due to the mastitis problems. On June 9, 2006 the largest milk processor in the world and the two largest supermarkets in the United States—Dean Foods, Wal-Mart, and Kroger—announced that they are "on a nationwide search for rBGH-free milk." No study has indicated that consumption of rBST-produced milk increases IGF1 levels, nor has any study demonstrated an increased risk of any disease between those consuming rBST and non-rBST produced milk. In 1994, the FDA stated that no significant difference has been shown between milk derived from rBST-treated and non-rBST-treated cows.

Milk may contain varying levels of white blood cells depending upon the health of the source animals, according to guidelines set up by the Food and Drug Administration and statistics reported by the dairy industry Although not considered a human health issue by most authorities, elevated white blood cell levels indicate an immune response by cattle, due in part to mastitis. There are concerns regarding the transmission of bovine paratubeculosis through somatic cells to humans but the evidence is largely inconclusive.

LACTOSE INTOLERANCE

Lactose, the disaccharide sugar component of all milk must be cleaved in the small intestine by the enzyme lactase in order for its constituents (galactose and glucose) to be absorbed. The production of this enzyme declines significantly after weaning in all mammals including humans (except for most northern westerners and a few other ethnic groups, lactase decline occurs after weaning, sometime between the ages of two and five). Once lactase levels have decreased sufficiently, consumption of small amounts of lactose can cause diarrhea, intestinal gas, cramps and bloating, as the undigested lactose travels through the gastrointestinal tract and serves as nourishment for intestinal microflora who excrete gas, a process known as anaerobic respiration.

Nutrition — Comparison by Animal Source

Milk composition analysis, per 100 grams

Constituents	*Unit*	*Cow*	*Goat*	*Sheep*	*Buffalo*
Water	g	87.8	88.9	83.0	81.1
Protein	g	3.2	3.1	5.4	4.5
Fat	g	3.9	3.5	6.0	8.0
Carbohydrate	g	4.8	4.4	5.1	4.9
Energy	kcal	66	60	95	110
	kJ	275	253	396	463
Sugar (lactose)	g	4.8	4.4	5.1	4.9
Fatty acids:					
Saturated	g	2.4	2.3	3.8	4.2
Mono-unsaturated	g	1.1	0.8	1.5	1.7
Polyunsaturated	g	0.1	0.1	0.3	0.2
Cholesterol	mg	14	10	11	8
Calcium	IU	120	100	170	195

Source: McCane, Widdowson, Scherz, Kloos.

These compositions vary by breed, animal, and point in the lactation period. Jersey cows produce milk of about 5.2% fat, Zebu cows produce milk of about 4.7% fat, Brown Swiss cows produce milk of about 4.0% fat, and Holstein-Friesian cows produce milk of about 3.6% fat. The protein range for these four breeds is 3.3% to 3.9%, while the lactose range is 4.7% to 4.9%.

Milk fat percentages in all dairy breeds vary according to digestible fibre, starch and oil intakes, and can therefore be manipulated by dairy farmers' diet formulation strategies. Mastitis infection can cause fat levels to decline.

Varieties and Brands

Milk products are sold in a number of varieties based on types/degrees of:

- age (e.g., cheddar);
- additives (e.g., vitamins);
- coagulation (e.g., cottage cheese);
- farming method (e.g., organic, grass-fed).
- fat content (e.g., half and half);
- fermentation (e.g., buttermilk);
- flavoring (e.g., chocolate);
- homogenization (e.g., raw milk);
- mammal (e.g., cow, goat, sheep);
- packaging (e.g., bottle);
- sterilization (e.g., pasteurization);
- water content (e.g., dry milk).

Organic Milk (in the United States) or Bio-Milk & Biologique Milk (in Europe) is milk produced without the use of chemical herbicides or pesticides, and generally with more natural fertilizers and higher standards for the animals, and is now easy to find on the shelves in many areas. Demeter certified milk is produced with biodynamic agriculture methods and is similar in standards to organic milk and biological milk, with a few special farm procedures added that are biodynamic-specific.

Cow milk is generally available in several varieties according to approximate butterfat content. See fat content of milk.

ADDITIVES AND FLAVORING

In countries where the cattle (and often the people) live indoors, commercially sold milk commonly has vitamin D added to it to make up for lack of exposure to UVB radiation. Reduced fat milks often have added vitamin A to compensate for the loss of the vitamin during fat removal; in the United States this results in reduced fat milks having a higher vitamin A content than whole milk.

To aid digestion in those with lactose intolerance, milk is available in some areas with added bacterial cultures such as *Lactobacillus acidophilus* ("acidophilus milk") and bifidobacteria ("a/B milk"). Another milk with *Lactococcus lactis* bacteria cultures ("cultured buttermilk") is often used in cooking to replace the traditional use of naturally soured milk, which has become rare due to the ubiquity of pasteurization which kills the naturally occurring lactococcus bacteria. Milk often has flavoring added to it for better taste or as a means of improving sales. Chocolate flavored milk has been sold for many years and has been followed more recently by such other flavors as strawberry and banana. South Australia has the highest consumption of flavored milk per person in the world, where Farmers Union Iced Coffee outsells Coca-Cola, a success shared only by Inca Kola in Peru and Irn-Bru in Scotland. Switzerland has a soft drink based on milk that tastes and looks much like SevenUp. This popular "milk-cola", named Rivella, is in fact the national soft drink and even comes complete in low calore and low sugar varieties. In spite of what might be expected, it does not taste like milk.

Distribution

Because milk spoils so easily, it should, ideally, be distributed as quickly as possible. In many countries milk used to be delivered to households daily, but economic pressure has made milk delivery much less popular, and in many areas daily delivery is no longer available. People buy it chilled at grocery or convenience stores or similar retail outlets. Prior to the widespread use of platics, milk was sold in wax-coated paper containers; prior to that mlk was often distributed to consumers in glass bottles; and before glass bottles, in bulk that was ladled into the customer's container.

In the UK, milk can be delivered daily by a milkman who travels his local milk round (route) using a milk float (often battery powered) during the early hours. Milk is delivered in 1 pint glass bottles with aluminium foil tops. Silver top denotes full cream unhomogenized; red top full cream homogenized; red/silver top semi-skimmed; blue/silver check top skimmed; and gold top channel island.

Empty bottles are rinsed before being left outside for the milkman to collect and take back to the dairy for washing and reuse. Currently many milkmen operate franchises as opposed to being employed by the dairy and payment is made at regular intervals, by leaving a check; by cash collection; or direct debit.

Although there was a steep decline in doorstep delivery sales throughout the 1990s, the service is still prominent, as dairies have diversified and the service is becoming more popular again. The doorstep delivery of milk is seen as part of the UK's heritage, and is relied upon by people up and down the country.

In New Zealand, milk is no longer distributed in glass bottles. In rural India, milk is delivered daily by a local milkman carrying bulk quantities in a metal container, usually on a bicycle; and in other parts of metropolitan India, milk is usually bought or delivered in a plastic bags or cartons via-shops or supermarkets.

In the United States, glass milk bottles have been mostly replaced with milk cartons (tall paper boxes with a square cross-section and a peaked top that can be folded outward upon opening to form a spout) and plastic jugs. Gallons of milk are almost always sold in jugs, while half-gallons and quarts may be found in both paper cartons and plastic jugs, and smaller sizes are almost always in cartons. Recently, milk has been sold in smaller resealable bottles made to fit in automobile cup holders. These individual serving sizes are also sold in flavored varieties.

The half-pint milk carton is the traditional unit as a component of school lunches. In the U.S., pictures of missing children were printed on the larger milk cartons as a public service until it was determined that this was disturbing to children.

Milk preserved by the UHT process is sold in cartons often called a brick that lack the peak of the traditional milk carton. Milk preserved in this fashion does not need to be refrigerated before opening and has a longer shelf life than milk in ordinary packaging. It is more typically sold unrefrigerated on the shelves in Europe and Latin America than in the United States.

Glass milk containers are now rare. Most people purchase milk in bags, plastic jugs or plastic-coated paper cartons. Ultraviolet (UV) light from fluorescent lighting can alter the flavor of milk, so many companies that once distributed milk in transparent or highly translucent containers are now using thicker materials that block the UV light. Many people feel that such "UV protected" milk tastes better.

MILK COMES IN A VARIETY OF CONTAINERS WITH LOCAL VARIANTS

- *Australia and New Zealand:* Distributed in a variety of sizes, most commonly in aseptic cartons for up to 1 litres, and plastic screw-top bottles beyond that with the following volumes; 1.1L, 2L, and 3L. 1 litre Bags are starting to appear in supermarkets, but have not yet proved popular. Most UHT-milk is packed in 1 or 2 litre paper containers with a sealed plastic spout.
- *Brazil:* Used to be sold in cooled 1 litre bags, just like in South Africa. Nowadays the most common form is 1 litre aseptic cartons containing UHT skimmed, semi-skimmed or whole milk, although the plastic bags are still in use.

- *Canada:* 1.33 litre plastic bags (sold as 4 litres in 3 bags) are widely available in some areas (especially the Maritimes, Ontario and Québec), although the 4 litre plastic jug has supplanted them in western Canada. Other common packaging sizes are 2 litre, 1 litre, 500 millilitre, and 250 millilitre cartons, as well as 4 litre, 1 litre, 250 mL aseptic cartons and 500 millilitre plastic jugs.
- *China:* Sweetened milk is a drink popular with students of all ages and is often sold in small plastic bags complete with straw. Adults not wishing to drink at a banquet often drink milk served from cartons or milk tea.
- *Parts of Europe:* Sizes of 500 millilitres, 1 litre (the most common), 2 litres and 3 litres are commonplace.
- *Hong Kong* - milk is sold in glass bottles (220 mL), cartons (236 mL and 1L), plastic jugs (2 litres) and aseptic cartons (250 mL).
- *India:* Commonly sold in 500 mL plastic bags. It is still customary to serve the milk boiled, despite pasteurization. Milk is often buffalo milk. Flavored milk is sold in most convenience stores in waxed cardboard containers. Convenience stores also sell many varieties of milk (such as flavored and ultra-pasteurized) in different sizes, usually in aseptic cartons.
- *Indonesia:* Usually sold in 1 litre cartons, but smaller, snack-sized cartons are available.
- *Israel:* Non-UHT milk is most commonly sold in 1 litre waxed cardboard boxes and 1 litre plastic bags. It may also be found in 0.5L and 2L waxed cardboard boxes, 2L plastic jugs and 1L plastic bottles. UHT milk is available in 1 litre (and less commonly also in 0.25L) carton “bricks”.
- *Japan:* Commonly sold in 1 litre waxed cardboard boxes. In most city centers there is also home delivery of milk in glass jugs. As seen in China, sweetened and flavored milk drinks are commonly seen in vending machines.
- *South Africa:* Commonly sold in 1 litre bags. The bag is then placed in a plastic jug and the corner cut off before the milk is poured.

- *South Korea:* sold in cartons (180mL, 200mL, 500mL 900mL, 1L, 1.8L, 2.3L), plastic jugs (100Ml and 1.8L), aseptic cartons (180mL and 200mL) and plastic bags (100mL).
- *Poland:* UHT milk is mostly sold in aseptic cartons (500mL, 1L, 2L), and non-UHT in 1L plastic bags or plastic bottles. Milk, UHT is commonly boiled, despite being pasteurized.
- *Turkey:* Commonly sold in 500 mL or 1L cartons or special plastic bottles. UHT milk is more popular. Milkmen also serve in smaller towns and villages.
- *United Kingdom:* Most stores still stock Imperial sizes: 1 pint (568 mL), 2 pints (1.136 L), 4 pints (2.273 L), 6 pints (3.408 L) or a combination including both metric and imperial sizes. Glass milk bottles delivered to the doorstep by the milkman are typically pint-sized and are returned empty by the householder for repeated reuse. Milk is also sold at supermarkets in either aseptic cartons or HDPE bottles. Milk can still be legally sold by the Imperial pint in reusable bottles in the UK under EU regulations (a distinction only shared with beer and cider), whilst a growing number of manufacturers such as Northern Foods now sell milk in 1 and 2 litre bottles.
- *United States:* Commonly sold in gallon, half-gallon and quart containers (U.S. customary units) of rigid plastic or, occasionally for sizes less than a gallon, waxed cardboard, although bottles made of opaque PET are starting to become more commonplace in all smaller sizes. The US single-serving size is usually the half-pint (about 240 ml). Occasionally dairies will deliver milk straight to customers in coolers filled with glass bottles (usually half-gallon). Some convenience store chains in the United States (such as Kwik Trip in the Midwest) sell milk in 1/2 gallon bags.
- *Uruguay:* Commonly sold in 1 litre bags. The bag is then placed in a plastic jug and the corner cut off before the milk is poured.

Practically everywhere, condensed milk and evaporated milk is distributed in metal cans, 250 and 125 ml paper containers and 100 and 200 mL squeeze tubes, and powdered milk (skim and whole) is distributed in boxes or bags.

Spoilage and Fermented Milk Products

When raw milk is left standing for a while, it turns "sour". This is the result of fermentation, where lactic acid bacteria ferment the lactose inside the milk into lactic acid. Prolonged fermentation may render the milk unpleasant to consume. This fermentation process is exploited by the introduction of bacterial cultures (e.g. *Lactobacilli* sp., *Streptococcus* sp., Leuconostoc sp., etc) to produce a variety of fermented milk products. The reduced pH from lactic acid accumulation denatures proteins and caused the milk to undergo a variety of different transformations in appearance and texture, ranging from an aggregate to smooth consistency. Some of these products include sour cream, yoghurt, cheese, buttermilk, viili, kefir and kumis. See Dairy product for more information.

Pasteurization of cow milk initially destroys any potential pathogens and increases the shelf-life, but eventually results in spoilage that makes it unsuitable for consumption. This causes it to assume an unpleasant odor, and the milk is deemed non-consumable due to unpleasant taste and an increased risk of food poisoning. In raw milk, the presence of lactic acid-producing bacteria, under suitable conditions, ferments the lactose present to lactic acid. The increasing acidity in turn prevents the growth of other organisms, or slows their growth significantly. During pasteurization however, these lactic acid bacteria are mostly destroyed.

In order to prevent spoilage, milk can be kept refrigerated and stored between 1 and 4 degrees Celsius in bulk tanks. Most milk is pasteurized by heating briefly and then refrigerated to allow transport from factory farms to local markets. The spoilage of milk can be forestalled by using ultra-high temperature (UHT) treatment; milk so treated can be stored unrefrigerated for several months until opened. Sterilized milk, which is heated for a much longer period of time, will last even longer, but also loses more nutrients and assume a different taste. Condensed milk, made by removing most of the water, can be stored in cans for many years, unrefrigerated, as can evaporated milk. The most durable form of milk is milk powder, which is produced from milk by removing almost all water. The moisture content is usually less than five percent in both drum and spray dried milk powder.

Language and Culture

The importance of milk in human culture is attested to by the numerous expressions embedded in our languages, for example "the milk of human kindness". In ancient Greek mythology, the goddess Hera spilled her breast milk after refusing to feed Heracles, resulting in the Milky Way.

In African and Asian developing nations, butter is traditionally made from fermented milk rather than cream. It can take several hours of churning to produce workable butter grains from fermented milk.

Holy books have also mentioned milk; the Bible contains references to the Land of Milk and Honey. In the Quran, there is a request to wonder on milk as follows: 'And surely in the livestock there is a lesson for you, We give you to drink of that which is in their bellies from the midst of digested food and blood, pure milk palatable for the drinkers.'(16-The Honeybee, 66). The Ramadhan fast is traditionally broken with a glass of milk and dates.

The verb, "to milk" something is often used in the vernacular of many English-speaking countries as a synonym for extortion or, in less loaded terms, taking advantage of a situation where one has another person at a disadvantage.

The word milk has had many slang meanings over time. In the early 17th century the word was used to mean semen, or vaginal secretions, or to masturbate oneself or someone else. In the 19th century, milk was used to describe a cheap alcoholic drink made from methylated spirits mixed with water. The word was also used to mean defraud, to be idle, to intercept telegrams addressed to someone else, and a weakling. In the mid 1930s, the word was used in Australia meaning to siphon gas from a car.

Milk is sometimes referred to as moo juice in American English, while Cockney rhyming slang calls it Acker Bilk, Tom Silk, Lady in silk and Kilroy silk.

The name of the Russian Molokan (Russian: "religion in Russian is derived from Russian " meaning "Milk" as they they would drink milk on the Russian Orthodox days of fast.

FAT CONTENT OF MILK

The fat content of milk is the proportion of milk made up by butterfat. The fat content, particularly of cow's milk, is modified to make a variety of products. The fat content of milk is usually stated on the container, and the colour of the label or milk bottle top varied to enable quick recognition.

Methods for Changing Fat Content

To reduce the fat content of milk, e.g. for skimmed or semi-skimmed milk, all of the fat is removed and then the required quantity returned. The fat content of the milk produced by cows can also be altered, by selective breeding and genetic modification. For example, scientists in New Zealand have bred cows that produce skimmed milk (less than 1% fat content

Methods of Detecting Fat Content

Milk's fat content can be determined by experimental means, such as the Babcock test or Gerber Method. Before the Babcock test was created, dishonest milk dealers could adulterate milk to falsely indicate a higher fat content. In 1911, the American Dairy Science Association's Committee on Official Methods of Testing Milk and Cream for Butterfat met in Washington DC with the U.S. Bureau of Dairying, the U.S. Bureau of Standards and manufacturers of glassware. Standard specifications for the Babcock methodology and equipment were published as a result of this meeting.[3] Improvements to the Babock test have continued.

Terms for Fat Content by Country

The terminology for different types of milk, and the regulations regarding labelling, varies by country and region.

Canada

In Canada "whole" milk refers to creamline (unhomogenized) milk. "Homogenized" milk refers to milk which is 3.25% butterfat. Generally all store-bought milk in Canada has been homogenized. Yet, the term is also used as a name to describe butterfat content for a specific variety of milk. Modern commercial dairy processing techniques involve first removing all of the butterfat, and then adding back the appropriate amount depending on which product is being produced on that particular line.

In the U.S. and Canada, a blended mixture of half cream and half milk is often sold in small quantities and is called half-and-half. Half-and-half is used for creaming coffee and similar uses. In Canada, low-fat cream is available, which has half the fat content of half-and-half.

United Kingdom

Three main varieties of milk by fat content are sold in the UK, skimmed, semi-skimmed and whole milk. These make up 17%, 58% and 25% of the market respectively Until 1 January 2008, milk with butterfat content outside the ranges defined by the European Commission could not legally be sold as milk. This included 1% milk, meaning The One, a 1% variety launched by Robert Wiseman Dairies, could not be labelled as milk. Lobbying by Britain has allowed these other percentages to be sold as milk. Since the change in regulation, Sainsbury's has launched a 1% variety with an orange milk bottle top.

10

LACTATION

INTRODUCTION

Lactation describes the secretion of milk from the mammary glands, the process of providing that milk to the young, and the period of time that a mother lactates to feed her young. The process occurs in all female mammals, and in humans it is commonly referred to as breastfeeding or nursing. In most species milk comes out of the mother's nipples; however, the platypus (a non-placental mammal) releases milk through ducts in its abdomen. In only one species of mammal, the Dayak fruit bat, is milk production a normal male function. In some other mammals, the male may produce milk as the result of a hormone imbalance. This phenomenon may also be observed in newborn infants as well (for instance witch's milk).

Galactopoiesis is the maintenance of milk production. This stage requires Prolactin (PRL) and Oxytocin.

Purpose

The chief function of lactation is to provide nutrition to the young after birth. In almost all mammals, lactation, or more correctly the suckling stimulus, induces a period of infertility, usually by the suppression of ovulation, which serves to provide the optimal birth spacing for survival of the offspring.

HUMAN LACTATION

Hormonal Influences

From the fourth month of pregnancy (the second and third trimesters), a woman's body produces hormones that stimulate the growth of the milk duct system in the breasts:

- *Progesterone* — influences the growth in size of alveoli and lobes. Progesterone levels drop after birth. This triggers the onset of copious milk production.
- *Oestrogen* — stimulates the milk duct system to grow and become specific. Oestrogen levels also drop at delivery and remain low for the first several months of breastfeeding. It is recommended that breastfeeding mothers avoid oestrogen-based birth control methods, as a spike in estrogen levels may reduce a mother's milk supply.
- Follicle stimulating hormone (FSH)
- Luteinizing hormone (LH)
- *Prolactin* — contributes to the increased growth of the alveoli during pregnancy.
- *Oxytocin* — contracts the smooth muscle of the uterus during and after birth, and during orgasm. After birth, oxytocin contracts the smooth muscle layer of band-like cells surrounding the alveoli to squeeze the newly-produced milk into the duct system. Oxytocin is necessary for the milk ejection reflex, or let-down to occur.
- *Human placental lactogen (HPL)* — From the second month of pregnancy, the placenta releases large amounts of HPL. This hormone appears to be instrumental in breast, nipple, and areola growth before birth.

By the fifth or sixth month of pregnancy, the breasts are ready to produce milk. It is also possible to induce lactation without pregnancy.

Lactogenesis I

During the latter part of pregnancy, the woman's breasts enter into the Lactogenesis I stage. This is when the breasts make colostrum (see below), a thick, sometimes yellowish fluid. At this stage, high levels of progesterone inhibit most milk production. It is not a medical concern if a pregnant woman leaks any colostrum before her baby's birth, nor is it an indication of future milk production.

Lactogenesis II

At birth, prolactin levels remain high, while the delivery of the placenta results in a sudden drop in progesterone, estrogen, and HPL levels. This abrupt withdrawal of progesterone in the presence of high prolactin levels stimulates the copious milk production of Lactogenesis II.

When the breast is stimulated, prolactin levels in the blood rise, peak in about 45 minutes, and return to the pre-breastfeeding state about three hours later. The release of prolactin triggers the cells in the alveoli to make milk. Prolactin also transfers to the breast milk. Some research indicates that prolactin in milk is higher at times of higher milk production, and lower when breasts are fuller, and that the highest levels tend to occur between 2 a.m. and 6 a.m

Other hormones—notably insulin, thyroxine, and cortisol—are also involved, but their roles are not yet well understood. Although biochemical markers indicate that Lactogenesis II begins about 30–40 hours after birth, mothers do not typically begin feeling increased breast fullness (the sensation of milk "coming in") until 50–73 hours (2–3 days) after birth.

Colostrum is the first milk a breastfed baby receives. It contains higher amounts of white blood cells and antibodies than mature milk, and is especially high in immunoglobulin A (IgA), which coats the lining of the baby's immature intestines, and helps to prevent germs from invading the baby's system. Secretory IgA also helps prevent food allergies. Over the first two weeks after the birth, colostrum production slowly gives way to mature breast milk.

Lactogenesis III

The hormonal endocrine control system drives milk production during pregnancy and the first few days after the birth. When the milk supply is more firmly established, autocrine (or local) control system begins. This stage is called Lactogenesis III

During this stage, the more that milk is removed from the breasts, the more the breast will produce milk. Research also suggests that draining the breasts more fully also increases the rate of milk production Thus the milk supply is strongly influenced by how often the baby feeds and how well it is able to transfer milk from the breast. Low supply can often be traced to:

- not feeding or pumping often enough;
- inability of the infant to transfer milk effectively caused by, among other things:
  - jaw or mouth structure deficits;
  - poor latching technique;
- rare maternal endocrine disorders;
- hypoplastic breast tissue;
- a metabolic or digestive inability in the infant, making it unable to digest the milk it receives;
- inadequate calorie intake or malnutrition of the mother;
- Lack of sexual activity. Frequent sexual activity in lactating women has been shown to increase milk supply by up to 25% in several clinical studies.

The release of the hormone oxytocin leads to the milk ejection or let-down reflex. Oxytocin stimulates the muscles surrounding the breast to squeeze out the milk. Breastfeeding mothers describe the sensation differently. Some feel a slight tingling, others feel immense amounts of pressure or slight pain/discomfort, and still others do not feel anything different.

The let-down reflex is not always consistent, especially at first. The thought of breastfeeding or the sound of any baby can stimulate this reflex, causing unwanted leakage, or both breasts may give out milk when an infant is feeding from one breast. However, this and other problems often settle after two weeks of feeding. Stress or anxiety can cause difficulties with breastfeeding.

A poor milk ejection reflex can be due to sore or cracked nipples, separation from the infant, a history of breast surgery, or tissue damage from prior breast trauma. If a mother has trouble breastfeeding, different methods of assisting the milk ejection reflex may help. These include feeding in a familiar and comfortable location, massage of the breast or back, or warming the breast with a cloth or shower.

Afterpains

The surge of oxytocin that may also possibly trigger the milk ejection reflex also causes the uterus to contract. During breastfeeding, mothers may feel these contractions as afterpains. These may range from period-like cramps to strong labour-like contractions and can be more severe with second and subsequent babies.

Lactation without Pregnancy, induced Lactation, Relactation

Milk production can be "artificially" and intentionally obtained in the absence of pregnancy in the woman. It is not necessary that the woman has ever been pregnant, and she can be well in her post-menopausal period. Women who have never been pregnant are sometimes able to induce enough lactation to breastfeed. This is called "induced lactation". Women who have breastfed before can be able to re-starts. This is called "relactation". If the nipples are consistently stimulated by a breast pump or actual suckling (several times a day), massaging and squeezing the female breasts or with additional help from temporary use of milk-inducing drugs, the breasts will eventually begin to produce enough milk to begin feeding a baby. Once established, lactation adjusts to demand. This is how some adoptive mothers, usually beginning with a supplemental nursing system or some other form of supplementation, can breastfeed. There is thought to be little or no difference in milk composition whether lactation is artificially induced or a result of pregnancy. A lactogene effect of herbs is not clinically confirmed, although several herbs are recommended to increase or evoke milk flow. These are for example fenugreek (the most popular), blessed thistle, and red raspberry leaf. Also, some couples may use lactation for sexual purposes. Rare accounts of male lactation (as distinct from galactorrhea) exist in medical literature.

11

CHEESE

Cheese is a food made from milk, usually the milk of cows, buffalo, goats, or sheep, by coagulation. The milk is acidified, typically with a bacterial culture, then the addition of the enzyme rennet or a substitute (e.g. acetic acid or vinegar) causes coagulation, to give "curds and whey". Some cheeses also have molds, either on the outer rind (similar to a fruit peel) or throughout.

Hundreds of types of cheese are produced. Their different styles, textures and flavors depend on the origin of the milk (including the animal's diet), whether it has been pasteurized, butterfat content, the species of bacteria and mold, and the processing including the length of aging. Herbs, spices, or wood smoke may be used as flavoring agents. The yellow to red color of many cheeses is a result of adding annatto. Cheeses are eaten both on their own and cooked in various dishes; most cheeses melt when heated.

For a few cheeses, the milk is curdled by adding acids such as vinegar or lemon juice. Most cheeses are acidified to a lesser degree by bacteria, which turn milk sugars into lactic acid, then the addition of rennet completes the curdling. Vegetarian alternatives to rennet are available; most are produced by fermentation of the fungus *Mucor miehei*, but others have been extracted from various species of the *Cynara* thistle family.

Cheese has served as a hedge against famine and is a good travel food. It is valuable for its portability, long life, and high content of fat, protein, calcium, and phosphorus. Cheese is more compact and has a longer shelf life than the milk from which it is made. Cheesemakers near a dairy region may benefit from fresher, lower-priced milk, and lower shipping costs. The long storage life of cheese allows selling it when markets are more favorable.

Etymology

The origin of the word cheese appears to be the Latin caseus, from which the modern word casein is closely derived. The earliest source is probably from the proto-Indo-European root *kwat-, which means "to ferment, become sour".

In the English language, the modern word cheese comes from chese (in Middle English) and ciese or cese (in Old English). Similar words are shared by other West Germanic languages — West Frisian *tsiis*, Dutch *kaas*, German *Käse*, Old High German *chasi* — all of which probably come from the reconstructed West-Germanic root kasjus, which in turn is an early borrowing from Latin.

The Latin word caseus is also the source from which are derived the Spanish queso, Portuguese queijo, Malay/ Indonesian Language keju (a borrowing from the Portuguese word queijo), Romanian cas and Italian cacio.

The Celtic root which gives the Irish cáis and the Welsh caws are also related.

When the Romans began to make hard cheeses for their legionaries' supplies, a new word started to be used: formaticum, from caseus formatus, or "molded cheese" (as in "formed", not "moulded"). It is from this word that we get the French *fromage*, Italian *formaggio*, Catalan *formatge*, Breton *fourmaj* and Provençal *furmo*. Cheese itself is occasionally employed in a sense that means "molded" or "formed". Head cheese uses the word in this sense.

Origins

Cheese is an ancient food whose origins predate recorded history. There is no conclusive evidence indicating where cheesemaking originated, either in Europe, Central Asia or the Middle East, but the practice had spread within Europe prior to Roman times and, according to Pliny the Elder, had become a sophisticated enterprise by the time the Roman Empire came into being.

Proposed dates for the origin of cheesemaking range from around 8000 BCE (when sheep were first domesticated) to around 3000 BCE. The first cheese may have been made by people in the Middle East or by nomadic Turkic tribes in Central Asia. Since animal skins and inflated internal organs have, since ancient times, provided storage vessels for a range of foodstuffs, it is probable that the process of cheese making was discovered accidentally by storing milk in a container made from the stomach of an animal, resulting in the milk being turned to curd and whey by the rennet from the stomach. There is a widely told legend about the discovery of cheese by an Arab trader who used this method of storing milk. The legend has many individual variations.

Dunlop cheese, a traditional cheese from Clerkland Farm, East Ayrshire, Scotland.Cheesemaking may also have begun independent of this by the pressing and salting of curdled milk in order to preserve it. Observation that the effect of making milk in an animal stomach gave more solid and better-textured curds, may have led to the deliberate addition of rennet.

The earliest archaeological evidence of cheesemaking has been found in Egyptian tomb murals, dating to about 2000 BCE. The earliest cheeses were likely to have been quite sour and salty, similar in texture to rustic cottage cheese or feta, a crumbly, flavorful Greek cheese.

Cheese produced in Europe, where climates are cooler than the Middle East, required less salt for preservation. With less salt and acidity, the cheese became a suitable environment for beneficial microbes and molds, giving aged cheeses their pronounced and interesting flavors.

Ancient Greece and Rome

Ancient Greek mythology credited Aristaeus with the discovery of cheese. Homer's Odyssey (8th century BCE) describes the Cyclops making and storing sheep's and goats' milk cheese. From Samuel Butler's translation:

> "We soon reached his cave, but he was out shepherding, so we went inside and took stock of all that we could see. His cheese-racks were loaded with cheeses, and he had more lambs and kids than his pens could hold...
>
> When he had so done he sat down and milked his ewes and goats, all in due course, and then let each of them have her own young. He curdled half the milk and set it aside in wicker strainers.

By Roman times, cheese was an everyday food and cheesemaking a mature art, not very different from what it is today. Columella's *De Re Rustica* (circa 65 CE) details a cheesemaking process involving rennet coagulation, pressing of the curd, salting, and aging. Pliny's *Natural History* (77 CE) devotes a chapter (XI, 97) to describing the diversity of cheeses enjoyed by Romans of the early Empire. He stated that the best cheeses came from the villages near Nîmes, but did not keep long and had to be eaten fresh. Cheeses of the Alps and Apennines were as remarkable for their variety then as now. A Ligurian cheese was noted for being made mostly from sheep's milk, and some cheeses produced nearby were stated to weigh as much as a thousand pounds each. Goats' milk cheese was a recent taste in Rome, improved over the "medicinal taste" of Gaul's similar cheeses by smoking. Of cheeses from overseas, Pliny preferred those of Bithynia in Asia Minor.

Post-classical Europe

Rome spread a uniform set of cheesemaking techniques throughout much of Europe, and introduced cheesemaking to areas without a previous history of it. As Rome declined and long-distance trade collapsed, cheese in Europe diversified further, with various locales developing their own distinctive cheesemaking traditions and products. The British Cheese Board claims that Britain has approximately 700 distinct local cheeses;France and Italy have perhaps 400 each. (A French proverb holds there is a different French cheese for every day of the year, and Charles de Gaulle once asked "how can you govern a country in which there are 246 kinds of cheese?" Still, the advancement of the cheese art in Europe was slow during the centuries after Rome's fall. Many of the cheeses we know best today were first recorded in the late Middle Ages or after— cheeses like cheddar around 1500 CE, Parmesan in 1597, Gouda in 1697, and Camembert in 1791.

In 1546, John Heywood wrote in Proverbes of John Heywood that "the moon is made of a *greene* cheese." (Greene may refer here not to the color, as many now think, but to being new or unaged.) Variations on this sentiment were long repeated. Although some people assumed that this was a serious belief in the era before space exploration, it is more likely that Heywood was indulging in nonsense.

Modern Era

Until its modern spread along with European culture, cheese was nearly unheard of in oriental cultures, uninvented in the pre-Columbian Americas, and of only limited use in sub-mediterranean Africa, mainly being widespread and popular only in Europe and areas influenced strongly by its cultures. But with the spread, first of European imperialism, and later of Euro-American culture and food, cheese has gradually become known and increasingly popular worldwide, though still rarely considered a part of local ethnic cuisines outside Europe, the Middle East, and the Americas.

The first factory for the industrial production of cheese opened in Switzerland in 1815, but it was in the United States where large-scale production first found real success. Credit usually goes to Jesse Williams, a dairy farmer from Rome, New York, who in 1851 started making cheese in an assembly-line fashion using the milk from neighboring farms. Within decades hundreds of such dairy associations existed.

The 1860s saw the beginnings of mass-produced rennet, and by the turn of the century scientists were producing pure microbial cultures. Before then, bacteria in cheesemaking had come from the environment or from recycling an earlier batch's whey; the pure cultures meant a more standardized cheese could be produced.

Factory-made cheese overtook traditional cheesemaking in the World War II era, and factories have been the source of most cheese in America and Europe ever since. Today, Americans buy more processed cheese than "real", factory-made or not.

MAKING CHEESE

Curdling

The only strictly required step in making any sort of cheese is separating the milk into solid curds and liquid whey. Usually this is done by acidifying (souring) the milk and adding rennet. The acidification is accomplished directly by the addition of an acid like vinegar in a few cases (paneer, queso fresco), but usually starter bacteria are employed instead. These starter bacteria convert milk sugars into lactic acid. The same bacteria (and the enzymes they produce) also play a large role in the eventual flavor of aged cheeses. Most cheeses are made with starter bacteria from the *Lactococci*, *Lactobacilli*, or *Streptococci* families. Swiss starter cultures also include Propionibacter shermani, which produces carbon dioxide gas bubbles during aging, giving Swiss cheese or Emmental its holes.

Some fresh cheeses are curdled only by acidity, but most cheeses also use rennet. Rennet sets the cheese into a strong and rubbery gel compared to the fragile curds produced by acidic coagulation alone. It also allows curdling at a lower acidity—important because flavor-making bacteria are inhibited in high-acidity environments. In general, softer, smaller, fresher cheeses are curdled with a greater proportion of acid to rennet than harder, larger, longer-aged varieties.

Curd Processing

At this point, the cheese has set into a very moist gel. Some soft cheeses are now essentially complete: they are drained, salted, and packaged. For most of the rest, the curd is cut into small cubes. This allows water to drain from the individual pieces of curd.

Some hard cheeses are then heated to temperatures in the range of 35 °C–55 °C (100 °F–130 °F). This forces more whey from the cut curd. It also changes the taste of the finished cheese, affecting both the bacterial culture and the milk chemistry. Cheeses that are heated to the higher temperatures are usually made with thermophilic starter bacteria which survive this step—either lactobacilli or streptococci.

Salt has a number of roles in cheese besides adding a salty flavor. It preserves cheese from spoiling, draws moisture from the curd, and firms up a cheese's texture in an interaction with its proteins. Some cheeses are salted from the outside with dry salt or brine washes. Most cheeses have the salt mixed directly into the curds.

A number of other techniques can be employed to influence the cheese's final texture and flavor. Some examples:

- *Stretching:* (Mozzarella, Provolone) The curd is stretched and kneaded in hot water, developing a stringy, fibrous body.
- *Cheddaring:* (Cheddar, other English cheeses) The cut curd is repeatedly piled up, pushing more moisture away. The curd is also mixed (or milled) for a long period of time, taking the sharp edges off the cut curd pieces and influencing the final product's texture.
- *Washing:* (Edam, Gouda, Colby) The curd is washed in warm water, lowering its acidity and making for a milder-tasting cheese.

Most cheeses achieve their final shape when the curds are pressed into a mold or form. The harder the cheese, the more pressure is applied. The pressure drives out moisture —

the molds are designed to allow water to escape — and unifies the curds into a single solid body.

Aging

A newborn cheese is usually salty yet bland in flavor and, for harder varieties, rubbery in texture. These qualities are sometimes enjoyed—cheese curds are eaten on their own—but normally cheeses are left to rest under carefully controlled conditions. This aging period (also called ripening, or, from the French, affinage) can last from a few days to several years. As a cheese ages, microbes and enzymes transform its texture and intensify its flavor. This transformation is largely a result of the breakdown of casein proteins and milkfat into a complex mix of amino acids, amines, and fatty acids.

Some cheeses have additional bacteria or molds intentionally introduced to them before or during aging. In traditional cheesemaking, these microbes might be already present in the air of the aging room; they are simply allowed to settle and grow on the stored cheeses. More often today, prepared cultures are used, giving more consistent results and putting fewer constraints on the environment where the cheese ages. These cheeses include soft ripened cheeses such as Brie and Camembert, blue cheeses such as Roquefort, Stilton, Gorgonzola, and rind-washed cheeses such as Limburger.

TYPES OF CHEESE

Factors in Categorization

Factors which are relevant to the categorization of cheeses include:

- Length of aging;
- Texture;
- Methods of making;
- Fat content;
- Kind of milk;
- Country/Region of Origin;

List of Common Categories

No one categorization scheme can capture all the diversity of the worlds cheeses. In practice, no single system is employed and diffeent factors are emphasised in describing different classes of cheeses. This typical list of cheese categories is from foodwriter Barbara Ensrud:

- Fresh
- Whey
- Pasta filata
- Semi-soft
- Semi-firm
- Hard
- Double and triple cream
- Soft-ripened
- Blue vein
- Goat or sheep
- Strong-smelling
- Processed.

Fresh, whey and stretched curd cheeses

The main factor in the categorization of these cheese is their age. Fresh cheeses without additional preservatives can spoil in a matter of days.

For these simplest cheeses, milk is curdled and drained, with little other processing. Examples include cottage cheese, Romanian Cas, Neufchâtel (the model for American-style cream cheese), and fresh goat's milk chèvre. Such cheeses are soft and spreadable, with a mild taste.

Whey cheeses are fresh cheeses made from the whey discarded while producing other cheeses. Provencal Brousse, Corsican Brocciu, Italian Ricotta, Romanian Urda, Greek Mizithra, and Norwegian Geitost are examples. Brocciu is mostly eaten fresh, and is as such a major ingredient in Corsican cuisine, but it can be aged too.

Traditional pasta filata cheeses such as Mozzarella also fall into the fresh cheese category. Fresh curds are stretched and kneaded in hot water to form a ball of Mozzarella, which in southern Italy is usually eaten within a few hours of being made. Stored in brine, it can be shipped, and is known worldwide for its use on pizzas. Other firm fresh cheeses include paneer and queso fresco.

Classed by Texture

Categorizing cheeses by firmness is a common but inexact practice. The lines between "soft", "semi-soft", "semi-hard", and "hard" are arbitrary, and many types of cheese are made in softer or firmer variations. The factor controlling the hardness of a cheese is its moisture content which is dependent on the pressure with which it is packed into molds and the length of time it is aged.

Semi-soft cheeses and the sub-group, Monastery cheeses have a high moisture content and tend to be bland in flavor. Some well-known varieties include Havarti, Munster and Port Salut.

Cheeses that range in texture from semi-soft to firm include Swiss-style cheeses like Emmental and Gruyère. The same bacteria that give such cheeses their holes also contribute to their aromatic and sharp flavors. Other semi-soft to firm cheeses include Gouda, Edam, Jarlsberg and Cantal. Cheeses of this type are ideal for melting and are used on toast for quick snacks.

Harder cheeses have a lower moisture content than softer cheeses. They are generally packed into molds under more pressure and aged for a longer time. Cheeses that are semi-hard to hard include the familiar Cheddar, originating in the village of Cheddar in England but now used as a generic term for this style of cheese, of which varieties are imitated worldwide and are marketed by strength or the length of time they have been aged. Cheddar is one of a family of semi-hard or hard cheeses (including Cheshire and Gloucester) whose curd is cut, gently heated, piled, and stirred before being pressed into forms. Colby and Monterey Jack are similar but milder cheeses; their curd is rinsed before it is pressed, washing away some acidity and calcium. A similar curd-washing takes place when making the Dutch cheeses Edam and Gouda.

Hard cheeses — "grating cheeses" such as Parmesan and Pecorino Romano — are quite firmly packed into large forms and aged for months or years.

Classed by Content

Some cheeses are categorized by the source of the milk used to produce them or by the added fat content of the milk from which they are produced. While most of the world's commercially available cheese is made from cows' milk, many parts of the world also produce cheese from goats and sheep, well-known examples being Roquefort, produced in France, and Pecorino Romano, produced in Italy, from ewes's milk. One farm in Sweden also produces cheese from moose's milk Sometimes cheeses of a similar style may be available made from milk of different sources, Feta style cheeses, for example, being made from goats' milk in Greece and of sheep and cows milk elsewhere.

Double cream cheeses are soft cheeses of cows' milk which are enriched with cream so that their fat content is 60% or, in the cae of triple creams, 75%.

Soft Ripened and Blue-vein

There are three main categories of cheese in which the presence of mold is a significant feature: soft ripened cheeses, washed rind cheees and blue cheeses.

Soft-ripened cheeses are those which begin firm and rather chalky in texture but are aged from the exterior inwards by exposing them to mold. The mold may be a velvety bloom of *Penicillium candida* or *P. camemberti* that forms a flexible white crust and contributes to the smooth, runny, or gooey textures and more intense flavors of these aged cheeses. Brie and Camembert, the most famous of these cheeses, are made by allowing white mold to grow on the outside of a soft cheese for a few days or weeks. Goats' milk cheeses are often treated in a similar manner, sometimes with white molds (Chèvre-Boîte) and sometimes with blue.

Washed-rind cheeses are soft in character and ripen inwards like those with white molds; however, they are treated differently. Washed rind cheeses are periodically cured in a solution of saltwater brine and other mold-bearing agents which may include beer, wine, brandy and spices, making their surfaces amenable to a class of bacteria *Brevibacterium linens* (the reddish-orange "smear bacteria") which impart pungent odors and distinctive flavors. Washed-rind cheeses can be soft (Limburger), semi-hard (Munster), or hard (Appenzeller). The same bacteria can also have some impact on cheeses that are simply ripened in humid conditions, like Camembert.

So-called Blue cheese is created by inoculating a cheese with *Penicillium roqueforti* or *Penicillium glaucum*. This is done while the cheese is still in the form of loosely pressed curds, and may be further enhanced by piercing a ripening block of cheese with skewers in an atmosphere in which the mold is prevalent. The mold grows within the cheese as it ages. These cheeses have distinct blue veins which gives them their name, and, often, assertive flavors. The molds may range from pale green to dark blue, and may be accompanied by white and crusty brown molds.Their texture can be soft or firm. Some of the most renowned cheeses are of this type, each with its own distinctive color, flavor, texture and smell. They include Roquefort, Gorgonzola, and Stilton.

PROCESSED CHEESES

Processed cheese is made from traditional cheese and emulsifying salts, often with the addition of milk, more salt, preservatives, and food coloring. It is inexpensive, consistent, and melts smoothly. It is sold packaged and either pre-sliced or unsliced, in a number of varieties. It is also available in spraycans.

Eating and Cooking

At refrigerator temperatures, the fat in a piece of cheese is as hard as unsoftened butter, and its protein structure is stiff as well. Flavor and odor compounds are less easily liberated when cold. For improvements in flavor and texture, it is widely advised that cheeses be allowed to warm up to room temperature before eating. If the cheese is further warmed, to 26–32 °C (80–90 °F), the fats will begin to "sweat out" as they go beyond soft to fully liquid.

At higher temperatures, most cheeses melt. Rennet-curdled cheeses have a gel-like protein matrix that is broken down by heat. When enough protein bonds are broken, the cheese itself turns from a solid to a viscous liquid. Soft, high-moisture cheeses will melt at around 55 °C (131 °F), while hard, low-moisture cheeses such as Parmesan remain solid until they reach about 82 °C (180 °F). Acid-set cheeses, including halloumi, paneer, some whey cheeses and many varieties of fresh goat cheese, have a protein structure that remains intact at high temperatures. When cooked, these cheeses just get firmer as water evaporates.

Some cheeses, like raclette, melt smoothly; many tend to become stringy or suffer from a separation of their fats. Many of these can be coaxed into melting smoothly in the presence of acids or starch. Fondue, with wine providing the acidity, is a good example of a smoothly melted cheese dish. Elastic stringiness is a quality that is sometimes enjoyed, in dishes including pizza and Welsh rarebit. Even a melted cheese eventually turns solid again, after enough moisture is cooked off. The saying "you can't melt cheese twice" (meaning "some things can only be done once") refers to the fact that oils leach out during the first melting and are gone, leaving the non-meltable solids behind.

As its temperature continues to rise, cheese will brown and eventually burn. Browned, partially burned cheese has a particular distinct flavor of its own and is frequently used in cooking (e.g., sprinkling atop items before baking them).

Health and Nutrition

In general, cheese supplies a great deal of calcium, protein, and phosphorus. A 30-gram (1.1 oz) serving of cheddar cheese contains about 7 grams (0.25 oz) of protein and 200 milligrams of calcium. Nutritionally, cheese is essentially concentrated milk: it takes about 200 grams (7.1 oz) of milk to provide that much protein, and 150 grams (5.3 oz) to equal the calcium.

Cheese potentially shares milk's nutritional disadvantages as well. The Center for Science in the Public Interest describes cheese as America's number one source of saturated fat, adding that the average American ate 30 lb (14 kg) of cheese in the year 2000, up from 11 lb (5 kg) in 1970. Their recommendation is to limit full-fat cheese consumption to 2 oz (57 g) a week. Whether cheese's highly saturated fat actually leads to an increased risk of heart disease is called into question when considering France and Greece, which lead the world in cheese eating (more than 14 oz/400 g a week per person, or over 45 lb/ 20 kg a year) yet have relatively low rates of heart disease. This seeming discrepancy is called the French Paradox; the higher rates of consumption of red wine in these countries is often invoked as at least a partial explanation.

Some studies claim to show that cheeses including Cheddar, Mozzarella, Swiss and American can help to prevent tooth decay. Several mechanisms for this protection have been proposed:

- The calcium, protein, and phosphorus in cheese may act to protect tooth enamel.
- Cheese increases saliva flow, washing away acids and sugars.
- Cheese may have an antibacterial effect in the mouth.

Controversy

Effect on Sleep

A study by the British Cheese Board in 2005 to determine the effect of cheese upon sleep and dreaming discovered that, contrary to the idea that cheese commonly causes nightmares, the effect of cheese upon sleep was positive. The majority of the two hundred people tested over a fortnight claimed beneficial results from consuming cheeses before going to bed, the cheese promoting good sleep. Six cheeses were tested and the findings were that the dreams produced were specific to the type of cheese. Although the apparent effects were in some cases described as colorful and vivid, or cryptic, none of the cheeses tested were found to induce nightmares. However, the six cheeses were all British. The results might be entirely different if a wider range of cheeses were tested. Cheese contains tryptophan, an amino acid that has been found to relieve stress and induce sleep.

Casein

Like other dairy products, cheese contains casein, a substance that when digested by humans breaks down into several chemicals, including casomorphine, an opioid peptide. In the early 1990s it was hypothesized that autism can be caused or aggravated by opioid peptides. Based on this hypothesis, diets that eliminate cheese and other dairy products are widely promoted. Studies supporting these claims have had significant flaws, so the data are inadequate to guide autism treatment recommendations.

Lactose

Cheese is often avoided by those who are lactose intolerant, but ripened cheeses like Cheddar contain only about 5% of the lactose found in whole milk, and aged cheeses contain almost none Nevertheless, people with severe lactose intolerance should avoid eating dairy cheese. As a natural product, the same kind of cheese may contain different amounts of lactose on different occasions, causing unexpected painful reactions. As an alternative, also for vegans, there is already a wide range of different soy cheese kinds available. Some people suffer reactions to amines found in cheese, particularly histamine and tyramine. Some aged cheeses contain significant concentrations of these amines, which can trigger symptoms mimicking an allergic reaction: headaches, rashes, and blood pressure elevations.

Pasteurization

A number of food safety agencies around the world have warned of the risks of raw-milk cheeses. The U.S. Food and Drug Administration states that soft raw-milk cheeses can cause "serious infectious diseases including listeriosis, brucellosis, salmonellosis and tuberculosis" It is U.S. law since 1944 that all raw-milk cheeses (including imports since 1951) must be aged at least 60 days. Australia has a wide ban on raw-milk cheeses as well, though in recent years exceptions have been made for Swiss Gruyère, Emmental and Sbrinz, and for French Roquefort. There is a trend for cheeses to be pasteurized even when not required by law.

Compulsory pasteurization is controversial. Pasteurization does change the flavor of cheeses, and unpasteurized cheeses are often considered to have better flavor, so there are reasons not to routinely pasteurize all cheeses. Some say that health concerns are overstated, pointing out that pasteurization of the milk used to make cheese does not ensure its safety in any case.[28] This is supported by statistics showing that in some European countries where young raw-milk cheeses may legally be sold, most cheese-related food poisoning incidents were traced to pasteurized cheses.

Pregnant women may face an additional risk from cheese; the U.S. Centers for Disease Control has warned pregnant women against eating soft-ripened cheeses and blue-veined cheeses, due to the listeria risk, which can cause miscarriage or harm to the fetus during birth.

World Production and Consumption

Greece is the world's largest (per capita) consumer of cheese, with 27.3 kg eaten by the average Greek. (Feta accounts for three-quarters of this consumption.) France is the second biggest consumer of cheese, with 24 kg by inhabitant. Emmental (used mainly as a cooking ingredient) and Camembert are the most common cheeses in France Italy is the third biggest consumer by person with 22.9 kg. In the U.S., the consumption of cheese is quickly increasing and has nearly tripled between 1970 and 2003. The consumption per person has reached, in 2003, 14.1 kg (31 pounds). Fior di latte (commonly known as mozzarella) is America's favorite cheese and accounts for nearly a third of its consumption, mainly because it is one of the main ingredients of pizza.

Cultural Attitudes

Although cheese is a vital source of nutrition in many regions of the world, and is extensively consumed in others, its use as a nutritional product is not universal. Cheese is rarely found in East Asian dishes, as lactose intolerance is relatively common in that part of the world and hence dairy products are rare. However, East Asian sentiment against cheese is not universal; cheese made from yaks' (*chhurpi*) or mares' milk is common on the Asian steppes; the national dish of Bhutan, ema datsi, is made from homemade cheese and hot peppers and cheese such as Rushan and Rubing in Yunnan, China is produced by several ethnic minority groups by either using goat's milk in the case of rubing or cow's milk in the case of rushan. Cheese consumption is increasing in China, with annual sales more than doubling from 1996 to 2003 (to a still small 30 million U.S. dollars a year). Certain kinds of Chinese preserved bean curd are sometimes misleadingly referred to in English as "Chinese cheese", because of their texture and strong flavour.

Strict followers of the dietary laws of Islam and Judaism must avoid cheeses made with rennet from animals not slaughtered in a manner adhering to halal or kosher laws.[38] Both faiths allow cheese made with vegetable-based rennet or with rennet made from animals that were processed in a halal or kosher manner. Many less-orthodox Jews also believe that rennet undergoes enough processing to change its nature entirely, and do not consider it to ever violate kosher law. (See Cheese and kashrut.) As cheese is a dairy food under kosher rules it cannot be eaten in the same meal with any meat.

Many vegetarians avoid any cheese made from animal-based rennet. Most widely available vegetarian cheeses are made using rennet produced by fermentation of the fungus Mucor miehei. Vegans and other dairy-avoiding vegetarians do not eat real cheese at all, but some vegetable-based cheese substitutes (usually soy-and almond-based) are available.

Even in cultures with long cheese traditions, it is not unusual to find people who perceive cheese - especially pungent-smelling or mold-bearing varieties such as Limburger or Roquefort - as unappetizing, unpalatable, or disgusting. Food-science writer Harold McGee proposes that cheese is such an acquired taste because it is produced through a process of controlled spoilage and many of the odor and flavor molecules in an aged cheese are the same found in rotten foods. He notes, "An aversion to the odor of decay has the obvious biological value of steering us away from possible food poisoning, so it is no wonder that an animal food that gives off whiffs of shoes and soil and the stable takes some getting used to."

Collecting Cheese Labels is called "Tyrosemiophilia"

In Language

In modern English slang, something "cheesy" is kitsch, cheap, inauthentic, or of poor quality. The use of the word probably derived not from the word cheese, but from the Persian or Hindi word *chiz*, meaning a thing. The word was picked up by British soldiers in Asia minor and came to mean "showy" in English slang from which it came to its modern usage.

A more whimsical bit of American and Canadian slang refers to school buses as "cheese wagons", a reference to school bus yellow. Subjects of photographs are often encouraged to "say cheese!", as the word "cheese" contains the phoneme /i/, a long vowel which requires the lips to be stretched in the appearance of a smile. People from Wisconsin and the Netherlands, both centers of cheese production, have been called cheeseheads. This nickname has been embraced by Wisconsin sports fans – especially fans of the Green Bay Packers or Wisconsin Badgers – who are often seen in the stands sporting plastic or foam hats in the shape of giant cheese wedges.

One can also be "cheesed off" – unhappy or annoyed. Also "Cheese it" is a 1950s slang term that means "get away fast".

12

Powdered Milk

Powdered milk is a manufactured dairy product made by evaporating milk to dryness. One purpose of drying milk is to preserve it; milk powder has a far longer shelf life than liquid milk and does not need to be refrigerated, due to its low moisture content. Another purpose is to reduce its bulk for economy of transportation. Available as Dry Whole Milk (DWM), it is most commonly produced as Non-Fat Dry Milk (NFDM), also known as Dried Skim Milk (DSM).

History and Manufacture

While Marco Polo wrote of Mongolian Tatar troops in the time of Kublai Kahn carrying sun-dried skimmed milk as "a kind of paste" the first usable commercial process to produce dried milk was invented by T.S. Grimwade and patented in 1855, though a William Newton had patented a vacuum drying process as early as 1837. Today, powdered milk is usually made by spray drying nonfat skim milk or whole milk. Pasteurized milk is first concentrated in an evaporator to about 50% milk solids. The resulting concentrated milk is sprayed into a heated chamber where the water almost instantly evaporates, leaving fine particles of powdered milk solids.

Alternatively, the milk can be dried by drum drying. Milk is applied as a thin film to the surface of a heated drum, and the dried milk solids are then scraped off. Powdered milk made this way tends to have a cooked flavor, due to caramelization caused by greater heat exposure.

Another process is freeze drying, which preserves many nutrients in milk, compared to drum drying.

The drying method and the heat treatment of the milk as it is processed alters the properties of the milk powder (for example, solubility in cold water, flavor, bulk density).

Uses

Powdered milk is frequently used in the manufacture of infant formula, confectionery such as chocolate and caramel, and in recipes for baked goods where adding liquid milk would render the product too thin. Powdered milk is also widely used in various sweets such as the famous Indian milk balls known as gulab jamun and popular Pakistani sweet delicacy (sprinkled with desiccated coconut) known as Chum chum (made with skim milk powder).

Powdered milk is also a common item in UN food aid supplies, fallout shelters, warehouses, and wherever fresh milk is not a viable option. It is widely used in many developing countries because of reduced transport and storage costs (reduced bulk and weight, no refrigerated vehicles). As with other dry foods, it is considered nonperishable, and is favored by survivalists, hikers, and others requiring nonperishable, easy-to-prepare food.

Reconstituting one cup of milk from powdered milk requires one cup of potable water and one-third cup of powdered milk.

Powdered milk is also used in western blots as a blocking buffer to prevent nonspecific protein interactions[5], and is referred to as Blotto.

Food and Health

Nutritional Value

Milk powders contain all twenty standard amino acids (the building blocks of proteins) and are high in soluble vitamins and minerals. According to USAID the typical average amounts of major nutrients in the unreconstituted milk are (by weight) 36% protein, 52% carbohydrates (predominantly lactose), calcium 1.3%, potassium 1.8%. Their milk powder is fortified with Vitamin A and D, 3000IU and 600IU respectively per 100g. Inappropriate storage conditions (high relative humidity and high ambient temperature) can significantly degrade the nutritive value of milk powder.

Oxysterols

Commercial milk powders are reported to contain, in general, very low levels of oxysterols (OS)[9]. However, compared to fresh milk (trace levels), powdered milk is higher in oxysterols (oxidized cholesterol), up to 30µg/g, yet significantly lower than powdered eggs (200µg/g)[10]. The OS free radicals have been suspected of being initiators of atherosclerotic plaques.

Adulteration

In the 2008 baby milk scandal in China, melamine adulterant was found in Sanlu infant formula, added to fool tests into reporting higher protein content. Thousands became ill and some children died after consuming the product.

13

Condensed Milk

INTRODUCTION

Condensed milk is highly shelf-stable milk product made by evaporating much of the water from milk to create a thick, syrupy liquid, and adding sugar to it before canning the milk in a sterilized can. When well made, condensed milk can last unopened on the shelf for up to two years. A related dairy offering *evaporated milk* is made in a similar fashion, but no sugar is added. Many grocery stores carry evaporated and condensed milk for use as beverages and in baking projects.

The inspiration for condensed milk came in the mid 1800s, when it was difficult to obtain milk in regions isolated from dairies. Milk also tended to be collected in unclean conditions, and it could sometimes be unsafe to drink. Gail Borden, a farmer, experimented with the idea of removing some of the water content from milk and adding a stabilizer before canning the result, in the hopes that he could develop a shelf-stable milk product which could be easily transported and solid. In 1864, Eagle Brand Consolidated Milk began to be produced.

Initially, condensed milk was intended to be used as a beverage, either consumed plain or thinned with water. People who did not have access to fresh milk would have used evaporated or condensed milk instead and many companies began to supplement their condensed milk, to ensure that it would be nutritionally valuable to consumers. As pasteurization and refrigeration became more widespread, fresh milk replaced condensed milk as a beverage, but some recipes still call for the product, and it can also be useful for camping trips.

To make condensed milk, high quality milk is first pasteurized to remove any potential contamination. The pasteurized milk is transferred to a sealed evaporator in a closed pipe system and subjected to high pressure, which lowers the boiling point of the milk. As a result, a lower heat can be used to remove as much as 60% of the water content of the milk, which is homogenized, stabilized sweetened, and canned in sterile containers. Sugar helps to fight bacteria, making condensed milk particularly shelf-stable.

Some consumers confuse condensed milk and evaporated milk, since the two products are very similar. As a general rule, evaporated milk is unsweetened, while condensed milk has been made with sugar. Some old recipes may specify the use of "sweetened condensed milk" to ensure that cooks do not use evaporated milk. Condensed milk is often used in beverages as well, adding a thick texture and sweet milky flavour to things like Thai iced tea.

Condensed milk, also known as sweetened condensed milk, is cow's milk from which water has been removed and to which sugar has been added, yielding a very thick, sweet product that can last for years without refrigeration if unopened. The two terms, condensed milk and sweetened condensed milk, have become synonymous; though there have been unsweetened condensed milk products, today these are uncommon. Condensed milk is used in numerous dessert dishes in many countries, especially in Russia.

A related product is evaporated milk, which has undergone a more complex process and which is not sweetened.

History

According to the writings of Marco Polo, the Tartars were able to condense milk. Ten pounds of milk paste was carried by each man who would mix the product with water. However, this probably refers to the soft Tartar curd which can be made into a drink ("airan") by diluting it, and therefore to fermented, not fresh milk concentrate.

Condensed milk was first developed in the United States in 1856 by Gail Borden, Jr. in reaction to difficulties in storing milk for more than a few hours. Before this development, milk could only be kept fresh for a short while and so was only available in the immediate vicinity of a cow. While returning from a trip to England in 1851, Borden was devastated by the death of several children, apparently due to poor milk from shipboard cows. With less than a year of schooling and following in a wake of failures both of his own and others, Borden was inspired by the vacuum pan he had seen used by Shakers to condense fruit juice and was at last able to reduce milk without scorching or curdling it Even then, his first two factories failed and only the third, in Wassaic, New York, produced a usable milk derivative that was long-lasting and needed no refrigeration.

Probably of equal importance for the future of milk were Borden's requirements (the "Dairyman's Ten Commandments") for farmers who wanted to sell him raw milk: they were required to wash udders before milking, keep barns swept clean, and scald and dry their strainers morning and night. By 1858 Borden's milk, sold as Eagle Brand, had gained a reputation for purity, durability and economy.

In 1864, Gail Borden's New York Condensed Milk Company constructed the New York Milk Condensery in Brewster, New York.This condensery was the largest and most advanced milk factory and was Borden's first commercially successful plant. Over 200 dairy farmers supplied 20,000 gallons of milk daily to the Brewster plant as demand was driven by the Civil War.

The U.S. government ordered huge amounts of it as a field ration for Union soldiers during the American Civil War. This was an extraordinary field ration for the nineteenth century: a typical 400 g (14 oz) can contains 1,300 calories, 30 g each of protein and fat, and more than 200 g of carbohydrate. Soldiers returning home from the Civil War soon spread the word. By the late 1860s, condensed milk was a major product. The first Canadian condensery was built at Truro, Nova Scotia, in 1871. In 1899, E.B. Stuart opened the first Pacific Coast Condensed Milk Company (later known as the Carnation Milk Products Company) plant in Kent, Washington. Unfortunately, the condensed milk market developed a bubble. Too many manufacturers chased too little demand. By 1912, stocks of condensed milk were large and the price dropped. Many condenseries went out of business. In 1911, Nestlé constructed the world's largest condensed milk plant in Dennington, Victoria, Australia.

In 1914, Professor Otto F Hunziker, head of Purdue University's dairy department, self-published Condensed milk and milk powder: prepared for the use of milk condenseries, dairy students and pure food departments. This text, along with additional work of Professor Hunziker and others involved with the American Dairy Science Association, standardized and improved condensery operations in the U.S. and internationally. Hunziker's book was republished in a seventh edition in October 2007 by Cartwright Press.

The first World War regenerated interest in, and a market for, condensed milk, primarily due to its storage and transportation benefits. In the U.S., the higher price for raw milk paid by condenseries created significant problems for the cheese industry.

Production

Raw milk is clarified and standardized, and then is heated to 85-90°C for several seconds. This heating destroys some microorganisms, decreases fat separation and inhibits oxidation. Some water is evaporated from the milk and sugar is added to approximately 45%. This sugar is what extends the shelf life of sweetened condensed milk. Sucrose increases the liquid's osmotic pressure, which prevents microorganism growth. The sweetened evaporated milk is cooled and lactose crystallization is induced.

Current Use

Condensed milk is used in recipes for the popular Portuguese candy brigadeiro in which condensed milk is the main ingredient (the most famous condensed milk in Brazil is called Moça I mô:ssa I and is made by Nestlé), lemon meringue pie, key lime pie, caramel candies and other desserts. In parts of Asia and Europe, sweetened condensed milk is the preferred milk to be added to coffee or sweetened tea. Many countries in South East Asia use condensed milk to flavour their coffee, cà phê s?a dá in Vietnamese. A popular treat in Asia is to put condensed milk on toast and eat it in a similar way as jam and toast. Nestlé has even produced a squeeze bottle similar to Smucker's jam squeeze bottles for this very purpose. In New Orleans, it is commonly used as a topping on top of a chocolate or similar cream flavor snowball. In Scotland, it is used for a popular sweet called Tablet or Swiss Milk Tablet.

Substitutions

To gain condensed milk from 1 cup of evaporated milk one has to add 1¼ cups of sugar and dissolve it by heating the milk.

14

Evaporated Milk

Evaporated milk, also known as dehydrated milk, is a shelf-stable canned milk product with about 60% of the water removed from fresh milk. It differs from condensed milk, which contains added sugar. Condensed milk requires less processing since the added sugar inhibits bacterial growth.

When mixed with an equal amount of water, evaporated milk becomes the equivalent of fresh milk. This means the actual liquid product takes up half the space of fresh milk, making it attractive for shipping purposes and can a shelf life of months or even years, depending upon the brand. This made evaporated milk very popular before refrigeration as a safe and reliable substitute for perishable fresh milk, that could be shipped easily to locations lacking the means of safe milk production or storage. Households in the affluent Western World use it most often today for desserts and baking due to its unique flavor.

Definition

According to the U.S. Government (21CFR131.130): "Evaporated milk is the liquid food obtained by partial removal of water only from milk. It contains not less than 6.5 percent by weight of milk fat, not less than 16.5 percent by weight of milk solids not fat, and not less than 23 percent by weight of total milk solids ... It is homogenized. It is sealed in a container ... processed by heat ... to prevent spoilage."

Vitamin D: Each fluid ounce of the food shall contain 25 International Units (IU)

Vitamin A: is optional, but if added, each fluid ounce of the food shall contain not less than 125 IU.

History

Condensed milk was introduced to the U.S. by Gail Borden which he made using a process under the patent issued on August 19, 1856. It became popular for those people who were remote from farm sources, since it was capable of long term storage. The invention of evaporated milk followed three decades later when John B. Meyenberg emigrated to the U.S. from Switzerland where he had devised the process, but had no support to begin production. He obtained two U.S. patents for his process and sterilizing apparatus, issued on November 25, 1884. He formed the Helvetia Milk Condensing Company on February 14, 1885, with a number of farmers and businessmen of Highland, Illinois, as stockholders. By June 14, 1885, the first canned "Highland Evaporated Cream" was ready to be marketed.

There were problems with the new product, with premature spoilage in early batches. Over the next few years, Louis Latzer and Dr. Werner Schmidt solved the problems which had been found to be caused by bacteria. With the marketing efforts of John Wilde, the company became successful as Pet, Inc., and is now part of General Mills.

John P. Meyenberg, son of John B. Meyenberg, was the first American to evaporate goat's milk. He started the Meyenberg business in 1934, supplying goat milk products that are more digestible than cow's milk, and an alternative for people (like himself) who were allergic to cow's milk.

Modern Production Process

Evaporated milk is fresh, homogenized milk from which 60 percent of the water has been removed. It is then chilled, fortified with vitamins and stabilizers, packaged and sterilized. A slightly caramelized flavor results from the high heat process, and it is slightly darker in color than fresh milk. The evaporation process also concentrates the nutrients and the food energy. Thus, for the same weight, undiluted evaporated milk contains more food energy than fresh milk.

International

In Malaysia, evaporated milk contains palm oil. It is one of the ingredients to make Teh Tarik in Malaysia and Singapore. Also it is added in brewed tea and coffee to make Teh See and Kopi C respectively.

Notable Producers

Evaporated milk is sold by several manufacturers:

- Carnation Evaporated Milk or Nestlé Evaporated Milk by Nestlé. During the 1950s, it sponsored such classic television shows as "Burns & Allen" and "The Jack Benny Show", among others, and commercials for the milk were often worked into the episodes, which was then a common practice. It was sold by Carnation under the slogan "the milk from Contented Cows."

Evaporated milk is milk which has had about sixty per cent by the water removed via evaporation. It is then homogenized, rapidly chilled, fortified with vitamins and stabilizers, packaged, and finally sterilized. Standards require whole evaporated milk contain at least 7.9 per cent milk fat and 25.5 per cent milk solids. The high heat process gives it a bit of a caramelizef flavour, and it is slightly darker in colour than fresh milk. The evaporation process naturally concenrates the nutrients and the calories, so evaporated version are more calorie-laden and nutritious than their fresh counterparts. You'll find skim, low-fat and whole milk varieties of evaporated milk. Low-fat and skim versions are also required to have added vitamin A, while all have added vitamins D and C.

Calories in Milk

Milk skimmed (pasteurised)	34	0.1g
Semi-skimmed	47	1.6g
Whole milk full fat	66	3.9g
Condensed Milk whole	320	9.5g
Dried skimmed milk	360	0.6g
Evaporated milk whole	**155**	**9g**
Goats milk	62	3.5g
Sheeps milk	99	6g
Soya milk	35	2g
Half cream fresh	150	13g
Single cream	200	19g
Soured cream	209	20g
Whipping cream	370	40g
Double cream	450	49g
Clotted cream	590	62g
Cheese Calories	400	36g

15

Bovine Somatotropin

INTRODUCTION

Bovine somatotropin (abbreviated bST and BST) is a protein hormone produced in the pituitary glands of cattle. It is also called bovine growth hormone, or BGH.

BST can be produced synthetically, using recombinant DNA technology. The resulting product is called recombinant bovine somatotropin (rBST), recombinant bovine growth hormone (rBGH), or artificial growth hormone. It is administered to the cow by injection and used to increase milk production. Currently Monsanto is the only company that markets recombinant bovine somatotropin, under the trade name Posilac.

Physiology

A cow's pituitary gland naturally secretes BST into the bloodstream. Some of it latches on to receptors in the liver, which then produce Insulin-like Growth Factor 1 (IGF-1), which enters the blood as well. These two hormones have many different effects in the body, including increasing the breakdown of fat for energy and helping to prevent mammary cell death The combination of increased energy from increased fat breakdown and decrease in mammary cell death is thought to be the cause of higher milk production.

Studies have shown that there is no increase in the amount of BST secreted in the milk when a cow is injected with supplemental rBST. However, the studies have been conflicting about whether or not IGF-1 and IGF-2 output increases. The amount of IGF-1 and IGF-2 secreted varies greatly by stage of lactation and whether or not an animal is pregnant; most studies have shown that, while these hormones are slightly elevated overall, it falls within the normal range of variation.

Posilac

In 1937, the administration of BST was shown to increase the milk yield in lactating cows by preventing mammary cell death in dairy cattle. Until the 1980s, there was very limited use of the compound in agriculture as the sole source of the hormone was from bovine cadavers. During this time, the knowledge of the structure and function of the hormone increased. Monsanto developed a recombinant version of BST, brand-named Posilac, in 1994, which is produced through a genetically-engineered *E. coli*. A gene that codes for the sequence of amino acids that make up BST is inserted into the DNA of the E. coli bacterium. The bacteria are then broken up and separated from the rBST, which, then, is purified to produce the injectable hormone. Growth hormones associated with injections given to dairy cows to increase milk production are known under an assortment of terms, but these terms, in general, refer to the Monsanto product. The Monsanto fact sheet on its proprietary product states that, when injected into dairy cattle, the product can increase milk production by an average of more than 10% over the span of 300 days

In the second half of 2008, Elanco Animal Health, a division of Eli Lilly and Company, bought the rights to Posilac from Monsanto.

USE OF POSILAC

Posilac prevents mammary cell death in dairy cattle. As such, it does not increase milk production on a day-to-day basis, but rather prevents milk production from decreasing over the long term, thus resulting in higher overall production during a lactation. Because a cow's milk production increases and decreases during her lactation based upon a known curve, application of Posilac can be carefully planned to maximize results.

An average dairy cow begins her lactation with a moderate daily level of milk production. This daily output increases until, at about 70 days into the lactation, production peaks. From that time until the cow is dry, production slowly decreases. This increase and decrease in production is partially caused by the count of milk-producing cells in the udder. Cell counts begin at a moderate number, increase during the first part of the lactation, then decrease and the lactation proceeds. Once lost, these cells generally do not regrow until the next lactation.

To apply Posilac for maximum effect, farmers are recommended to make the first Posilac application about 50 days into the cow's lactation, just before she peaks. The Posilac then sustains already-present mammary cells, limiting the rate of production decrease after production peaks. After the peak, production declines with or without application of Posilac, but declines more slowly with Posilac than without. This decrease in the rate of production decline permits dairy cows to produce more milk over the span of a lactation - at its best, this will be seen by seven to eight more pounds of milk being produced per day than would be produced without the benefit of Posilac.

Controversy

Use of rBST is controversial because of its potential effects on animal and human health and the perceived encroachment on small farmers by large corporations.

Animal Health

Two meta-analyses have been published on rBST's effects on bovine health. Findings indicated an average increase in milk output ranging from 11%-16%, a nearly 25% increase in the risk of clinical mastitis, a 40% reduction in fertility and 55% increased risk of developing clinical signs of lameness.

The same study reported a decrease in body condition score but speculated that it may have been attributable to differences in feeding of treated (underfed) versus untreated (overfed) cows.

A European Union scientific commission was asked to report on the incidence of mastitis and other disorders in dairy cows and on other aspects of the welfare of dairy cows. The commission's statement, subsequently adopted by the European Union, stated that the use of rBST substantially increased health problems with cows, including foot problems, mastitis and injection site reactions, impinged on the welfare of the animals and caused reproductive disorders. The report concluded that, on the basis of the health and welfare of the animals, rBST should not be used. Health Canada prohibited the sale of rBST in 1999; the recommendations of external committees were that, despite not finding a significant health risk to humans, the drug presents a threat to animal health, and, for this reason, cannot be sold in Canada

Human Health

According to the FDA, the "overwhelming scientific opinion" is that rBST is safe for human consumption, and that no significant difference exists between milk derived from rBST-treated and non-rBST-treated cows. In 1990, an independent panel convened by the National Institute of Health reaffirmed the FDA opinion that milk and meat from cows supplemented with rBST is safe for human consumption.

Still, various consumer groups have expressed concern over perceived effects from both BST itself, as well as insulin-like growth factor 1 (IGF-1), which is increased by rBST injections. Monsanto has stated that both of these compounds are harmless given the levels found in milk and the effects of pasteurization.

Lawsuit Against Fox Television

Fox television affiliate WTVT/Fox13 in Tampa, Florida was sued by Steve Wilson and Jane Akre, two former employees who were fired in relation to their report on BST. The journalists originally wrote a story in 1996 that covered the potential for human health risks of rBST. The station began publicizing the upcoming broadcast of the story, but, after a threatening letter from Monsanto, the station asked for changes to the story. The journalists, however, refused to change the story, and, succumbing to threats to be reported to the FCC, the station fired the journalists. This story is featured at length in the documentaries The Corporation and Outfoxed.

After a five-week trial, Akre was awarded $425,000 in damages; Wilson was awarded nothing. The jury found that Fox's actions were in retaliation for Akre's refusal of "a false, distorted, or slanted story," in the words of the juryThe jury did not, however, agree that the station bowed to pressure from Monsanto to alter their reporting.

Fox appealed and prevailed February 14, 2003, when an appeals court issued a ruling reversing the jury. The court's basis was that FCC policies on news agencies reporting the truth are not legally binding; and, as such, Fox had no legal requirement to report the truth in a news story.

In 2004, Fox filed a $1.7 million counter-suit against Akre and Wilson for trial fees and costs.

IGF-1

Monsanto's studies show use of rBST in cows increases bovine insulin-like growth factor 1 in milka structure that is identical in cows and humans. Monsanto states that there is no danger of consuming milk or meat from cows treated by BST, and that the only difference between milk from supplemented cattle and unsupplemented cattle is the amount of IGF-1, though even these elevated levels are similar to levels found in milk from untreated cows. Further, the amount of IGF-1 consumed in milk is negligible compared to the amount produced in the body.

Regulation

Use of the recombinant supplement has been controversial. While it is used in the United States (though not without reaction), it is banned in Canada, parts of the European Union, Australia and New Zealand.

Regulation inside the United States

In 1993, the product was approved for use in the U.S. by the FDA, and its use began in 1994. The product is now sold in 49 states, all but Michigan. According to Monsanto, approximately one-third of dairy cattle in the U.S. are treated with Posilac; approximately 8,000 dairy producers use the product. It is now the top-selling dairy cattle pharmaceutical product in the U.S.

Enforcement

The FDA does not require special labels for products produced from cows given rBST but has charged several dairies with "misbranding" their milk as having no hormones, because all milk contains hormones and cannot be produced in such a way that it would not contain any hormones. Monsanto sued an independent dairy over their use of a label which pledged to not use artificial growth hormones. The dairy stated that their disagreement was not over the scientific evidence for the safety of Posilac (Monsanto's complaint about the label), but rather they were more interested in marketing milk than a drug. The suit was settled when the dairy agreed to add a qualifying statement to their previous label regarding the lack of difference between milk produced by Posilac-dosed cows and that produced by cows that had not received the drug.

Demand for milk without using synthetic hormones has increased 500% in the US since Monsanto introduced their rBST product; organic milk is the fastest-growing sector of the organic food market.

Labelling in Pennsylvania

In 2007, the U.S. State of Pennsylvania adopted a regulation that would have banned the practice of labeling milk as derived from cows not treated with rBST. This prohibition was to go into effect January 1, 2008, but was delayed to February 1, 2008 in order to give interested parties more time to submit comments to the state's Department of Agriculture. This policy, had it been implemented, would have prevented consumers from distinguishing between milk from cows treated with rBST and milk from untreated cows. The ban was opposed by several consumer groups, and the state reversed its position before the ban could take effect, and adopted the Federal Trade.

Response from Commercial Groups

Several milk purchasers and resellers have elected not to purchase milk produced with rBST. The nation's largest dairy processor, Dean Foods, no longer sells milk from rBST-treated cows, and the top 3 grocery retailers in the nation, Wal-Mart, Kroger, and Costco have pledged not to sell such milk in their stores. Specific examples include:

- Winder Farms, a home delivery dairy and grocer in Utah and Nevada, sells rBST-free milk.
- Safeway in the northwestern United States stopped buying from dairy farmers that use rBST in January 2007. The two Safeway plants produce milk for all of Oregon, Southwest Washington, and parts of northern California. Safeway's plant in San Leandro, CA had already been rBST-free for two years.
- Chipotle Mexican Grill has also announced it will serve rBST-free sour cream at its restaurants.
- Kroger has banned rBST-derived milk from all its stores (including its subsidiaries such as Ralphs) as of February 2008.
- Publix announced it has been rBST-free since May, 2007.
- Braum's has also issued a press release stating its milk is rBST-free.
- Starbucks Company has as of January 2008 made all dairy in beverages rBST free.
- Wal-Mart and Sam's Club stores featured hormone-free "Great Value" brand milk, but did not label it as such in 2008.

Monsanto has responded to this trend by lobbying state governments to ban the practice of distinguishing between milk from farms pledged not to use rBST and those that do. According to the New York Times a pro-rBST advocacy group called Afact has been most active in these lobbying efforts. Afact is made up of both dairy farmers and allied industries, and is closely affiliated with Monsanto itself; the group's acronym stands for American Farmers for the Advancement and Conservation of Technology. Though rBST is one of Afact's main concerns, their mission is to prevent "marketers from convincing some consumers to doubt the credibility and safety assurances from of even the most respected food safety agencies and scientific oversight organizations."

Thus far, a large-scale negative consumer response to Afact's legislative and regulatory efforts has kept state regulators from pushing through strictures that would ban hormone-free milk labels, though several politicians have tried, including Pennsylvania's (see the Pennsylvania section above) agriculture secretary Dick Wolff, who tried to ban rBST-free milk on the grounds that it would alleviate consumer confusion. Proposed labeling changes have been floated by Afact lobbyists in New Jersey, Ohio, Indiana, Kansas, Utah, Missouri and Vermont. So far, however, this effort has been unsuccessful.

Regulation Outside the United States

In Japan, Australia, New Zealand, and Canada, rBST is not approved for use.

The European Union declared the use of rBST as safe in 1990, but, in 1993, a moratorium was placed on its sale by all member nations. It was turned into a permanent ban starting from January 1, 2000.

Canada's health board, Health Canada, refused to approve rBST for use on Canadian dairies, citing concerns over animal health. The study they had commissioned, however, found "no biologically plausible reason for concern about human safety if rbST were to be approved for sale in Canada. The only exception to this statement is the occurrence of an antibody reaction (possible hypersensitivity) in a subchronic (90-day) study of rbST oral toxicity in rats that resulted in one test animal's developing an antibody response at low dose (0.1 mg/kg/day) after 14 weeks.

The Codex Alimentarius Commission, United Nations body that sets international food standards, has to date refused to approve rBST as safe. The Codex Alimentarius does not have authority to ban or approve the hormone but its decisions are regarded as a standard and approval by the Codex would have allowed exporting countries to challenge countries with a ban on rBGH before the WTO.

BOVINE MILK IN HUMAN NUTRITION

Milk and milk products are nutritious food items containing numerous essential nutrients, but in the western societies the consumption of milk has decreased partly due to claimed negative health effects. The content of oleic acid, conjugated linoleic acid, omega-3 fatty acids, short- and medium chain fatty acids, vitamins, minerals and bioactive compounds may promote positive health effects. Full-fat milk has been shown to increase the mean gastric emptying time compared to half-skimmed milk, thereby increasing the gastrointestinal transit time. Also the low pH in fermented milk may delay the gastric emptying. Hence, it may be suggested that ingesting full-fat milk or fermented milk might be favourable for glycaemic (and appetite?) regulation. For some persons milk proteins, fat and milk sugar may be of health concern. The interaction between carbohydrates (both natural milk sugar and added sugar) and protein in milk exposed to heat may give products, whose effects on health should be further studied, and the increasing use of sweetened milk products should be questioned. The concentration in milk of several nutrients can be manipulated through feeding regimes. There is no evidence that moderate intake of milk fat gives increased risk of diseases.

Bovine milk and dairy products have long traditions in human nutrition. The significance of milk is reflected in our northern mythology where a cow named Audhumla was evolved from the melting ice. She had horn and milk was running as rivers from her teats. This milk was the food for Ymer, the first creature ever existing.

The consumption of milk and milk products vary considerably among regions; of drinking milk from about 180 kg yearly per capita in Island and Finland to less than 50 kg in Japan and China. In the western societies, the consumption of milk has decreased during the last decades This trend may partly be explained by the claimed negative health effects that have been attributed to milk and milk products. This criticism has arisen especially because milk fat contains a high fraction of saturated fatty acids assumed to contribute to heart diseases, weight gain and obesity.

The association between food and health is well established and recent studies have shown that modifiable risk factors seem to be of greater significance for health than previously anticipated. Prevention of disease may in the future be just as important as treatment of diseases. Indeed, many consumers of today are highly aware of health-properties of food, and the market for healthy food and food with special health benefits is increasing.

Milk is a complex food made up of components, which per se may have negative or positive health effects, respectively. Milk composition can be altered by the feeding regime. The main aim of this review is to discuss effects of milk components that are of particular interest for human health, and to give an overview of the potential for manipulation of bovine milk by feeding regimes to the lactating cows, thus giving improved nutritional composition of the milk for human consumption.

MILK COMPOSITION IN GENERAL

Bovine milk contains the nutrients needed for growth and development of the calf, and is a resource of lipids, proteins, amino acids, vitamins and minerals. It contains immunoglobulins, hormones, growth factors, cytokines, nucleotides, peptides, polyamines, enzymes and other bioactive peptides. The lipids in milk are emulsified in globules coated with membranes. The proteins are in colloidal dispersions as micelles. The casein micelles occur as colloidal complexes of protein and salts, primarily calcium Lactose and most minerals are in solution. Milk composition has a dynamic nature, and the composition varies with stage of lactation, age, breed, nutrition, energy balance and health status of the udder. Colostrums differ considerably to milk; the most significant difference is the concentration of milk protein that may be about the double in colostrum compared to later in lactation he change in milk composition during the whole lactation period seems to match the changing need of the growing infant, giving different amounts of components important for nutrient supply, specific and non-specific host defence, growth and development. Specific milk proteins are involved in the early development of immune response, whereas others take part in the non-immunological defence (e.g. lactoferrin). Milk contains many different types of fatty acids All these components make milk a nutrient rich food item.

Components in Milk and their Health Effects

- Lipids
- Fatty acids

Lipids

In average, milk contains about 33 g total lipid (fat)/l . Triacylglycerols, which account for about 95% of the lipid fraction, are composed of fatty acids of different length (4–24 C-atoms) and saturation Each triacylglycerol molecule is built with a fatty acid combination giving the molecule liquid form at body temperature. Other milk lipids are diacylglycerol (about 2% of the lipid fraction), cholesterol (less than 0.5%), phospholipids (about 1%), and free fatty acids (FFA) accounting to less than 0.5% of total milk lipids . Increased levels of FFA in milk might result in off-flavours in milk and dairy products, and the free volatile short-chain fatty acids contribute to the characteristic flavours of ripened cheese.

Saturated Fatty Acids

More than half of the milk fatty acids are saturated, accounting to about 19 g/l whole milk . The specific health effects of individual fatty acids have been extensively studied Butyric acid is a well-known modulator of gene function, and may also play a role in cancer prevention Caprylic and capric acids (8:0 and 10:0) may have antiviral activities, and caprylic acid has been reported to delay tumour growth Lauric acid (12:0) may have antiviral and antibacterial functions and might act as an anti caries and anti plaque agent Interestingly, Helicobacter pylori can in fact be killed by this fatty acid Another interesting observation is that capric and lauric acid are reported to inhibit COX-I and COX-II Stearic acid (18:0) does not seem to increase serum cholesterol concentration, and is not atherogenic.

It would appear, accordingly, that some of the saturated fatty acids in milk have neutral or even positive effects on health. In contrast to this, the saturated fatty acids lauric-, myristic (14:0) and palmitic (16:0) acid have low-density lipoprotein (LDL) and high-density lipoprotein (HDL) cholesterol-increasing properties]. High intake of these acids raises blood cholesterol levels and diets rich in saturated fat have been regarded to contribute to development of heart diseases, weight gain and obesity Association between consumption of milk and milk products and serum total cholesterol, LDL cholesterol and HDL cholesterol has been reported . High cholesterol levels are a risk factor for coronary heart disease (CHD), with LDL cholesterol and a high ratio between LDL and HDL cholesterol enhancing the risk of CHD.

Several intervention studies have shown that diets containing low-fat dairy products have been associated with favourable changes in serum cholesterol However, milk fat consumption has been shown to have less pronounced effects on serum lipids than could be expected from the fat content To our knowledge epidemiological cohort studies does not show a higher risk for diseases in persons with high intakes of dairy fat, as also shown by Elwood et al. cohort studies provide no convincing evidence that milk is harmful. On the contrary, several studies have found a lack of association between milk consumption and CHD. Two Swedish studies have shown that cardiovascular risk factors were negatively associated with intake of milk fat A Norwegian study suggests that intake of dairy fat or some other component of dairy products, as reflected by C15:0 as marker in adipose tissue may protect persons at increased risk from having a first myocardial infarction (MI), and that the causal effects may rely on other factors than serum cholesterol. It has been shown that 34 grams dairy fat per day gives no negative effect on odds ratio for myocardial infarction. As reported by Sjogren et al., fatty acids typically found in milk products were associated with a more favourable LDL profile in healthy men (i.e., fewer small, dense LDL particles), and they concluded that men with high intakes of milk products had an apparently beneficial and reduced distribution of the harmful small, dense LDL particles.

A Canadian 13-year follow up study analysed plasma LDL sub fractions with different density, and showed that cardiovascular risk was largely related to accumulation of small, dense LDL particles The small, dense LDL particles are also reported to be associated with hypertriglyceridemia insulin resistance the metabolic syndrome and increased risk for CHD Saturated fatty acids increase the serum concentration of both LDL- and HDL cholesterol. In a metaanalysis of 60 selected trials Mensink et al.reported that saturated fatty acids gave an unchanged ratio between total cholesterol and HDL cholesterol if carbohydrates replaced saturated fatty acids. It was shown by Hostmark et al.that an index reflecting the LDL/HDL balance, ATH-index = (total cholesterol-HDL)*apoB/(apoA*HDL), improved the discrimination between controls and subjects with coronary artery stenosis. Unlike this, the distribution of total cholesterol was similar in controls and patients, as evaluated by coronary angiography. In keeping with these early results, in the INTERHEART case-control study on risk factors associated with myocardial infarction in 52 countries, an increase in apo B/apo A1 ratio was shown to be the strongest risk factor for myocardial infarctionApoB/apo A1 was found to be a stronger risk factor than total cholesterol alone, or ratio between LDL and HDL cholesterol (Yusuf, personal information).

Increased levels of C-reactive protein (CRP) have been associated with inflammation and CRP is recognized as a risk factor for CHD and metabolic syndrome. Fredrikson et al. found, however, no significant association between CRP and intake of saturated fat. These studies are in agreement with others.

The increase in HDL cholesterol caused by the saturated fatty acids lauric-, myristic- and palmitic acid has beneficial effects as the reverse cholesterol transport is increased HDL can also act as an antioxidant and prevent oxidation of LDL particles in the blood, and it may protect against infections and against toxins from microbes

UNSATURATED FATTY ACIDS

Oleic acid (18:1c9) is the single unsaturated fatty acid with the highest concentration in milk accounting to about 8 g/litre whole milk Accordingly milk and milk products contribute substantially to the dietary intake of oleic acid in many countries. In Norway about a quarter of the average intake of oleic acid comes from milk and milk product Oleic acid is considered to be favourable for health, as diets with high amounts of monounsaturated fatty acid will lower both plasma cholesterol, LDL-cholesterol and triacylglycerol concentrations and replacement of saturated fatty acids with cis-unsaturated fatty acids reduces risk for coronary artery disease Several studies also indicate a cancer protective effect of oleic acid, but the data are not fully convincing.

Fatty acids are the main building material of cell membranes. The unsaturated fatty acids are reactive as they may give oxidative stress with free radicals and secondary peroxidation products (different aldehydes such as malonedialdehyde and 4-hydroxynonenale) that may be harmful to proteins and DNA in the cells This may contribute to cancer and to mitochondrial aging processes caused by mutations in mitochondrial DNA The enzyme lechitin/cholesterol acyl transferace (LCAT), having an important role in reverse cholesterol transport, is sensitive to oxidative stress and it is also inhibited by minimally oxidized LDL Oleic acid is more stable to oxidation than the omega-3 and omega-6 fatty acids, and it can partly replace these fatty acids in both triacylglycerols and membranelipids. A high ratio between oleic acid and polyunsaturated fatty acids will protect lipids in i.e. LDL towards attack from oxidative stressors such as cigarette smoke, ozone and other oxidants. Studies have shown that a diet rich in monounsaturated/polyunsaturated fatty acids give better protection against atheromatosis and CVD than a diet rich in polyunsaturated fatty acids.

Milk fat is rich in oleic acid (about 25% oleic acid) and it has a very high ratio oleic acid/polyunsaturated fatty acids. A diet rich in milk fat therefore may help to increase this ratio in the total dietary fatty acids. A high intake of meat from i.e. sheep may be expected to have similar effect. This might partly explain why mortality by cardiac disease has been lower in Iceland compared to the other Scandinavian countries and the average age of living has been higher despite of higher intake of saturated fat (coming from both mutton and milk).

The concentration of PUFA in milk is about 2 g/land the main PUFA in milk are linoleic (18:2 omega-6) and alpha-linolenic (18:3 omega-3) acid . These fatty acids may be converted to fatty acids with 20 carbon atoms, i.e. arachidonic acid (20:4 omega-6) and eicosapentaenoic acid, (EPA) (20:5 omega-3), and further converted to eicosanoids; metabolically very active compounds with local functions. Eicosanoids derived from linoleic acid, via arachidonic acid, may enhance blood platelet aggregation and thereby increase the coronary risk, in contrary to eicosanoids produced form the long omega-3 fatty acids EPA has the ability to partially block the conversion of the omega-6 fatty acids to harmful eicosanoids, thereby reducing the cardiovascular risk and inhibiting tumour genesis. PUFA may also affect signal transduction and gene expression t is conceivable therefore that the type of fatty acid in the membrane governs several metabolic functions.

It has been argued that the Mesolithic man had a ratio of 1–4 : 1 between the omega-6 and omega-3 fatty acids, against now in most European diets 10–14 : 1 Eskimos, and some populations in Japan, having a high intake of omega-3 fatty acids, also have a low rate of coronary heart diseases, and of some cancers Conceivably, protection against cardiovascular diseases and cancer would be related to the ratio of EPA plus oleic acid to omega-6 fatty acids in the diet, and hence in the body.

In milk the ratio between omega-6 and omega-3- fatty acids is low and favourable compared to most other non-marine products This ratio is greatly influenced by the feeding regime, and may with favourable feeding be as low as 2:1. Comparing the omega-6 to omega-3 ratio in milk in the Nordic countries. Researchers have reported the lowest ratio in Iceland: 2.1:1, compared to 4.7:1 in milk from the other Nordic countries. It has been suggested that the higher supply of omega-3 fatty acids from milk in Iceland might explain the lower prevalence of type-2 diabetes and CHD mortality in Iceland compared to the other Nordic countries . A Norwegian study showed reduced risk of premenopausal breast cancer with milk intake With proper feeding regime, milk and meat from ruminants can in fact be the main source of omega-3 fatty acids in the human diet, as is the case in France

According to the above considerations a favourable meal should be rich in oleic acid, and have a low ratio between omega-6 fatty acids and omega-3 fatty acids, perhaps near 1–2:1. Indeed, milk fat fits into this description probably better that any other food item.

Conjugated linoleic acid (CLA)

Bovine milk, milk products and bovine meat are the main dietary sources of the cis9, trans 11 isomer of conjugated linoleic acid (9c,11t-CLA) In most cases this isomer is the most abundant CLA-isomer in bovine milk [64]. Minor amounts of other geometrical and positional isomers of CLA also occur in milk (such as the 7t, c9 and 10t, 12c-CLA), with different biological effects Milk content of 9c,11t-CLA vary considerably but may constitute about 0,6% of the fat fraction.

The health effects of CLA have been discussed Administration of 9c,11t CLA has shown to modulate plasma lipid concentration in both human and animal models. Some studies but not all have shown that addition of CLA isomer mixtures (9c,11t and 10t,12c) to a diet affects plasma lipids. Studies have shown that especially 9c,11t-CLA can improve plasma cholesterol status . In a study with healthy men Tricon et al.ound a significant reduction in plasma total cholesterol concentration by 9c,11t-CLA. The results concerning the effects of CLA on serum triglycerides are controversial Tricon et al. observed a decrease in serum triglycerides by 9c,11t-CLA compared to 10t,12c-CLA in humans, and Roche et al. found serum triglycerides and unesterified FA to be decreased by 9c,11t-CLA in ob/ob-mice.

In experimental animals CLA has been shown to have anticarcinogenic effects Prospective data from a Swedish study suggest that high intakes of high-fat dairy foods and CLA may reduce the risk of colorectal cancer The knowledge of CLA's effects in metabolism and the reported anti-proliferative and pro-apoptotic effect of CLA on various types of cancer cells makes CLA to an interesting, and possible therapeutic agent in nutritional cancer therapy. The mechanisms by which CLA might affect metabolism are many. It is suggested that CLA competes with arachidonic acid in the cyclooxygenase reaction, resulting in reduced concentration of prostaglandins and tromboxanes in the 2-series . CLA may suppress the gene expression of cyclooxygenase, and reduce the release of pro-inflammatory cytokines such as TNF-alpha and interleukines in animals. CLA also activates the PPARs transcription factors, and CLA may reduce the initial step in NF-kappa B activation and thereby reduce cytokines, adhesions molecules and other stress-induced molecules.

Trans Vaccenic Acid (VA)

The main trans 18:1 isomer in milk fat is vaccenic acid, (18:1, 11t, VA), but trans double bounds in position 4 to 16 is also observed in low concentrations in milk fat.

The amount of VA in milk fat may vary; constituting 1.7% or 4-6% of the total fatty acid content Typically, the concentration of VA may be about 2–4% when the cows are on fresh pasture and about 1–2% on indoor feeding Normally, naturally increase in 9c,11t-CLA in milk also results in increased concentration of VA.

The VA has a double role in metabolism as it is both a trans fatty acid and a precursor for 9c,11t-CLA. As demonstrated by Kay et al. approximately 90% of 9c,11t-CLA in milk fat was produced endogenously involving delta-9-desaturation of VA. Vaccenic acid can be converted to 9c,11t-CLA in rodents pigs and humans.

Trans fatty acids have been shown to increase blood lipids. Industrially produced trans fat are shown to increase the risk of coronary heart disease as they have adverse influence on the ratio of LDL on HDL, and on Lp(a)]. It has been questioned if VA has these same adverse effects. In one study with hamster, Meijer et al. found that VA was more detrimental to cardiovascular risk than elaidic acid (18:1, 9t) due to a more increasing effect on LDL/HDL cholesterol ratio. Furthermore, Clifton et al. showed that VA was an independent predictor of a first myocardial infarction. In contrast to this, it has been shown by Willett et al. that trans fat from animals did not give an increased risk for CHD. As recently demonstrated by Tricon et al a combination of naturally increased concentration of VA and 9c,11t-CLA in milk fat did not result in detrimental effects on most cardiovascular disease risk parameters. However, it remains to clarify if VA has unhealthy effects on blood lipids.

Phospholipids and Glycosphingolipids

Phospholipids and glycosphingolipids accounts to about one per cent of total milk lipids These lipids contain relatively larger quantities of polyunsaturated fatty acids than the triacylglycerols. They have functional roles in a number of reactions, such as binding cations, help to stabilize emulsions, affect enzymes on the globule surface, cell-cell interactions, differentiation, proliferation, immune recognition, transmembrane signalling and as receptors for certain hormones and growth factors Gangliosides are one of these components found in milk. Gangliosides (with more than one sialic acid moiety) are mainly found in nerve tissues, and they have been demonstrated to play important roles in neonatal brain development, receptor functions, allergies, for bacterial toxins etc.

Protein

Bovine milk contains about 32 g protein/l The milk protein has a high biological value, and milk is therefore a good source for essential amino acids. In addition, milk contains a wide array of proteins with biological activities ranging from antimicrobial ones to those facilitating absorption of nutrients, as well as acting as growth factors, hormones, enzymes, antibodies and immune stimulants . The nitrogen in milk is distributed among caseins, whey proteins and non-protein nitrogen. The casein content of milk represents about 80% of milk proteins. Caseins biological function is to carry calcium and phosphate and to form a clot in the stomach for efficient digestion. The milk whey proteins are globular proteins that are more water soluble than caseins, and the principle fractions are beta-lactoglobin, alpha-lactalbumin, bovine serum albumin and immunoglobulins. Whey is the liquid remaining after milk has been curdled to produce cheese, and it is used in many products for human consumption, such as ricotta and brown cheese, and concentrated whey is an additive to several products e.g. bread, crackers, pastry and animal feed. The rate at which the amino acids are released during digestion and absorbed into the circulation may differ among the milk proteins, and whey proteins are considered as rapid digested protein that gives high concentrations of amino acids in postprandial plasma . The benefit of drinking whey has been known for centuries, and two ancient proverbs from the Italian city of Florence say, “If you want to live a healthy and active life, drink whey” and, “If everyone was raised on whey, doctors would be bankrupt”.

Some of the milk proteins (e.g. secretory immunoglobulin A, lactoferrin, 1-antitrypsin, ß-casein and lactalbumin) may be relatively resistant to digestive enzymes, and the whole protein or peptides derived from it, may exert their function in the small intestine before being fully digested.

As several bioactive proteins and peptides derived from milk proteins are potential modulators of various regulatory processes in the body, some of these are produced on an industrial scale, and are considered for application as ingredients in both 'functional foods' and pharmaceutical preparations. Although the physiological significance of several of these substances is not yet fully understood, both the mineral binding and cytomodulatory peptides derived from bovine milk proteins are now claimed to be health enhancing components that can be used to reduce the risk of disease or to enhance a certain physiological function Milk protein composition may differ among breeds For example the concentration of beta-casein A1 is low in milk from cows in Iceland and in New Zealand. It has been speculated that this proteins may have a role in the development of diabetes and cardiac disease However, later it was concluded in a review article that there is no convincing evidence that the A1 beta-casein of cow milk has any adverse effect in humans.

MILK PEPTIDES AND BLOOD PRESSURE

Several studies has suggested that there is an association between milk consumption and blood pressure; as hypertension is inversely related to milk consumption in some epidemiological- and intervention studies It has been suggested that some milk peptides have antihypertensive effects, both by inhibiting angiotensin-converting enzyme, having opoid-like activities, antithrombotic properties and by binding minerals.

Milk is especially rich in essential amino acids and branched chain amino acids. There is evidence that these amino acids have unique roles in human metabolism; in addition to provide substrates for protein synthesis, suppress protein catabolism and serve as substrates for gluconeogenesis, they also trigger muscle protein synthesis and promote protein synthesis. Essential amino acids are shown to be more important than non-essential amino acids in muscle protein synthesisand the branched chain amino acid leucine in particular triggers muscle protein synthesis which is sensed by the insulin-signalling pathway. The stimulated insulin secretion caused by milk, is suggested to be caused by milk proteins, and as shown by Nilsson et al. a mixture of leucine, isoleucine, valine, lysine and threonine resulted in glycemic and insulinemic response resembling the response seen after ingestion of whey. A combination of milk with a meal with high glycaemic load (rapidly digested and absorbed carbohydrates) may stimulate insulin release and reduce the postprandial blood glucose concentrationA reduction in postprandial blood glucose is favourable, and it is epidemiological evidence suggesting that milk may lower risk of diseases related to insulin resistance syndrome.

TAURINE

The concentration of taurine is high in breast milk (about 18 mg/l) and in colostrum from cow, but in regular bovine milk it is not high; about 1 mg/l Goat milk is however very rich in taurine: 46–91 mg/l . Taurine is an essential amino acid for preterm neonates, and specific groups of individuals are at risk for taurine deficiency and may benefit from supplementation, e.g. patients requiring long-term parenteral nutrition (including premature and newborn infants); diabetes patients, those with chronic hepatic, heart or renal failure It is suggested that during parenteral nutrition, supplementation of 50 mg taurine per kg body weight may be required.

Taurine is the most abundant intracellular amino acid in humans. It may be synthesized in the body from methionine and cysteine, but in healthy individuals the diet is the usual source of taurine. It is implicated in numerous biological and physiological functions: bile acid conjugation and cholestasis prevention, antiarrhythmic/inotropic/chronotropic effects, central nervous system neuromodulation, retinal development and function, endocrine/metabolic effects and antioxidant/anti-inflammatory properties Taurine has been shown to have endothelial protective effects it may function principally as a negative feedback regulator, helping to dampen immunological reactions before they cause too much damage to host tissues or to the leukocytes themselves and it is shown to be analgesic.

Glutathione (GSH)

Fresh milk may be a good source of glutathione, a tripeptide of the sulphur amino acid cysteine, plus glycine and glutamic acid. In the organism glutathione has the role as an antioxidant. Glutathione can be oxidized forming GSSG (oxidized glutathione), and in this reaction it may remove reactive oxygenspecies (ROS), thereby regulating the level of ROS in the cells. Glutathione participates in regulation of insulin production in the pancreatic cells, as ROS inhibit expression of the pro insulin gene. Glutathione appears to have different important roles in leukocytes, as a growth factor, as an anti-apoptotic factor in leukocytes and to regulate the pattern of cytokine secretion. GSH, moreover, is also central for antioxidative defence in the lungs, which may be very important in connection with lower respiratory infections including influenza.

Minerals, Vitamins and Antioxidants

Milk contains many minerals, vitamins and antioxidants. The antioxidants have a role in prevention of oxidation of the milk, and they may also have protective effects in the milk-producing cell, and for the udder. Most important antioxidants in milk are the mineral selenium and the vitamins E and A. As there are many compounds that may have antioxidative function in milk, measurement of total antioxidative capacity of milk may be a useful tool

Calcium

The calcium concentration in bovine milk is about 1 g/l. Dairy products provide more than half of the calcium in the typical American diet and daily intake of milk and milk products thus has a central role in securing calcium intake. In human nutrition adequate calcium intake is essential. Getting enough calcium in the diet gives healthy bones and teeth, and it may also help prevent hypertension, decrease the odds of getting colon or breast cancer, improve weight control and reduce the risk of developing kidney stones.

Selenium

The selenium concentration in body fluids and tissues are directly related to selenium intake. The selenium concentration in Scandinavian food is low, and the concentration in Norwegian bovine milk is about 11 μg/l (own results, 2006), and 37 μg/l in the USFor plant products the situation is even worse; the selenium concentration in wheat flour (whole grain) is less than 20 ug/kg in Norwegian wheat (own results), compared to 707 ug/kg in the US.

Selenium is important in human health; it has a role in the immune- and antioxidant system and in DNA synthesis and DNA repair . Selenoprotein P is an antioxidative defence enzyme having similar function as the selenoenzyme phospholipid hydro peroxide glutathione peroxidase (Gpx-4) inside the cells and it also protects LDL towards peroxidation A strong negative correlation between the concentration of selenium and the concentration of plasma lipid peroxidation products has been reported in a Canadian study . These observation are in line with epidemiological observations from USA, where a strong negative correlation between mortality of ischemic cardiac disease and hypertension among men and women in the age group 55–64 years comparing states with different selenium intake.

Selenium protects against many (but not all) types of cancer There are indications that selenium may protect against asthma, and that low selenium intake may worsen the asthma symptoms Selenium deficiency has even been linked to adverse mood states Selenium is also a component of enzymes involved in metabolism of thyroid hormone.

As selenium is of fundamental importance to human health, the low selenium availability in Scandinavian soil is of concern. Different strategies can be used to increase human selenium intake, and addition of selenium-rich yeast to the feed of domestic animals is one option. Recommended daily intake of selenium is 55 μg, and the optimal selenium concentration in bovine milk may be discussed. If milk contains about 50–100 μg selenium/l, it would be a good selenium source.

Iodine

Iodine is an essential component of the thyroid hormones. These hormones control the regulation of body metabolic rate, temperature regulation, reproduction and growth.

The recommended iodine intake is 150 µg/d for adults. Accordingly, a daily intake of 0.5 litres milk with an average content of 160 µg iodine/l meets about 50% of the requirement. However, it is important to underline the great seasonal variation in iodine content of milk.

Magnesium

Magnesium is ubiquitous in foods, and milk is a good source, containing about 100 mg/l milkRecommended intake is 400 mg/day for men and 310 mg/day for women Magnesium has many functions in the body, participating in more than 300 reactions. Magnesium deficiency has been linked to atherosclerosis, as studies have shown that deficiency may give oxidative stress. Magnesium may also have a role in reducing asthma, and experimental studies of persons with asthma suggest that magnesium infusion may have a place in the acute treatment of asthma A possible mechanism may be that magnesium together with taurine dampens the signaling effects of a too high calcium release inside the cells Magnesium deficiency may occur following kidney disease and after use of some diuretic drugs. Magnesium deficiency in elderly is observed, and may be a result of poor appetite or unbalanced diet.

Zinc

Zinc is an essential part of several enzymes and metalloproteins. Zinc has several functions in the body, in DNA repair, cell growth and replication, gene expression, protein and lipid metabolism, immune function, hormone activity, etc Milk is a good zinc source; containing about 4 mg/l. Recommended intake is 8 and 11 mg/day for adult female and male The bioavailability of zinc is better from milk than from vegetable food and inclusion of milk in the diet may improve total bioavailability of zinc.

Vitamin A

Milk is a good source of retinoids, containing 280 µg/l Recommended daily intake is 700–900 µg/day Vitamin A has a role in vision, proper growth, reproduction, and immunity, cell differentiation, in maintaining healthy bones as well as skin and mucosal membranes.

Vitamin E

Vitamin E concentration in milk is about 0,6 mg/l, but may increase 3-4 folds by proper feeding regimes (see later). Recommended intake is 15 mg/day Vitamin E is not a single compound; it includes tocoferols and tocotrienols. In whole milk, alpha-tocopherol is the major form of vitamin E (>85%); gamma-tocopherol and alpha-tocotrienol are present to a lesser extent, about 4% each of the sum of tocoferols and tocotrienols. Observational studies indicate that high dietary intake of vitamin E are associated with decreased risk for cancer and coronary heart disease, and that vitamin E can stimulate T-cells and increase the immune defence system. Milk seems to be a food item favouring absorption and transportation of vitamin E from ingested food into the chylomicrons.

Vitamin B_{12}

Milk is also a good source of vitamin B_{12}, being 4.4 µg/l The daily recommendation is 2.4 µg Vitamin B_{12} is found only in animal foods, and plays a central role in folate and homocysteine metabolism, by transferring methyl groups. Vitamin B_{12} deficiency may cause megaloblastic anaemia and breakdown of the myelin sheath.

Folate

Bovine milk contains 50 µg folate/l Studies indicate that 5-methyl-tetrahydrofolate is the major folate form in milk Recommended intake of folate is 400 ug/day for adults Many scientists believe that folate deficiency is the most prevalent of all vitamin deficiencies It is generally accepted that folate supplementation (400 µg/day) before conception and during the first weeks of pregnancy reduces the risk of neural tube defects. A recent study has shown that higher total folate intake was associated with a decreased risk of incident hypertension, particularly in younger women In addition, folates may have a protective role to play against coronary heart disease and certain forms of cancer, but sufficient evidence is not yet available The complexity of the folate metabolism suggest that different metabolites of folate are involved in different reactions, and that dihydrofolate and 5-methyl-tetrahydrofolate are the active compounds in growth-inhibition in colon cancer cells.

The bioavailability of folate variesFolate-binding proteins occur in unprocessed milk, pasteurised milk, spray-dried skim milk powder and whey Animal and human studies have suggested that these components enhance food folate bioavailability, and it is shown that inclusion of cow milk in the diet enhanced the bioavailability of food folate In a population-based study, the consumption of milk and yogurt were inversely associated with serum total homocysteine concentrations, and the authors explained this association by intakes of folate and riboflavin.

Riboflavin

Milk is a good source of riboflavin, 1.83 mg riboflavin/l milk. Daily recommended intake is 1.1 and 1.3 mg for women and men, respectively Riboflavin is part of two important coenzymes participating in a numerous metabolic pathways in the cell. It has a role in the antioxidant performance of glutathione peroxidase and DNA repair via the ribonucleotid reductase pathway.

Bacterial Flora of Milk

Milk samples from normal healthy mammary glands contain many strains of bacteria. To prevent diseases caused by pathogenic bacteria in milk and to lengthen the shelf life of milk, treatment such as cooling and pasteurization or membrane filtration is needed. To preserve milk, addition of selective, well-documented strains of starter cultures for fermentation is a method that has been used for centuries.

16

Fermented Milk

Historically, the seasonal variation in milk production made it necessary to preserve milk. The Nordic countries including Iceland have a long tradition for using fermented milk, and the consumption of fermented milk is about 20 kg per person.

During fermentation bacteria and yeasts convert lactose in the milk to various degradation products depending on the species present. *Lactobacilli* and *streptococci* give rice to lactic acid and monosaccarides (especially galactose). Bifidobacteria give rice to lactic acid, acetic acid and monosaccarides, while yeasts, present only in some few fermented milk products, produce CO_2 and ethanol. Different bacterias may be used for fermentation, giving products of special flavour and aroma, and with several potential health beneficial metabolites The bacteria contain cell wall components that bind Toll-like receptors on dendritic cells (and also other leucocytes) found in the mucosa of the small intestine and colon, thus stimulating the Th1 immune response It has been shown that fermented milk stimulates the Th1 immune response, and down-regulates the Th2 immune response. The immune system may thus be strengthened against cancer, virus infections and allergy Bacterial DNA has also a similar effect, binding to Toll-like receptor-9 Some bacteria can also improve the intestinal microbial balance, and the fermented milk may have positive health effects both in the digestive channel and in metabolism. During the fermentation of milk, lactic acid and other organic acids are produced and these increase the absorption of iron. If fermented milk is consumed at mealtimes, these acids are likely to have a positive effect on the absorption of iron from other foods. Lactic acid is also a poorer substrate for growth of pathogenic bacteria than glucose and lactose.

The low pH in fermented milk may also delay the gastric emptying from the stomach into the small intestine and thereby increase the gastrointestinal transit time Also, full-fat milk has been shown to increase the mean gastric emptying half-time compared to half-skimmed milk and accordingly it might be favourable to gastric emptying and thus may have an effect on appetite regulation.

Intolerance to Milk Components

The public "belief" that milk causes an inflammatory process and an increase in mucus production has not been confirmed It has been shown that respiratory symptoms was not associated with milk intake and concluded that consumption of milk does not seem to exacerbate the symptoms of asthma, but in a few cases people with cow's milk allergy may have asthma-like symptoms after milk consumption . However in cells from another tissue; mucin producing cells of gastric mucosa, alpha-lactalbumin stimulates mucin synthesis and secretion.

Milk Allergy

Most milk proteins, even proteins present at low concentrations, are potential allergens. A person may be allergic to casein or whey proteins or to both. Milk allergy may arise in small children (0-3 years) and it is estimated that 2-5% of the children has milk allergy After the age of three years it is no longer a problem for most children.

Milk allergy reactions may either be of the type 'rapid onset' or the 'slower-onset' type. The rapid type comes suddenly with symptoms of e.g. wheezing, vomiting, anaphylaxis. The slower-onset reactions are more common and symptoms develop over a period of hours or days after ingesting milk, and may include loose stool, vomiting, fussiness, reduced weight gain etc. As these symptoms are more general, this type is difficult to diagnose.

Cow's milk allergy may be treated by completely avoiding milk proteins. Epitopes on milk proteins have been shown to be both conformational and linear epitopes, widely spread throughout the protein molecules. Due to the great variability and heterogeneity of the human IgE response, no single allergen or particular structure has been found to be a major part of milk allergenicity.

An interesting study from Germany showed that the children of farmers had less allergy, in spite of the fact that these children were drinking more whole- milk than other children not living on farms

Intolerance to Milk Proteins

There has been speculation if milk proteins may have a role in Attention Deficit Hyperactivity Disorder (ADHD), autism, depressions and schizophrenia in some cases. There are major supports to the hypothesis that ADHD may be linked to increased levels of neuroactive peptides and increased urinary peptide levels. A diet free of milk, milk products and gluten may in many cases give reduced ADHD symptoms. Further, opioid peptides derived from food proteins (exorphins) have been found in urine of autistic patients. This area of investigation is important and large scale, good quality randomised controlled trials are needed.

Lactose Intolerance

The lactose concentration in bovine milk is about 53 g/l. People often confuse a milk allergy with lactose intolerance, but they are not the same thing. Lactose intolerance is common in many adults throughout the world, and is caused by deficiency of intestinal lactase (hypolactasia). Lactose maldigestion occurs in about 75% of the worldwide population and about 25% of the US population. In Scandinavian counties it varies between 2% and 18%. Avoiding all lactose is seldom necessary, and persons with hypolactasia can usually ingest limited amounts of milk without having annoying symptoms. Individual differences in gut micro flora may be one reason for large variations in amounts of milk that is tolerated. To ingest milk with a meal may also improve tolerance. Instead of drinking regular milk, fermented milk may be an option, because fermented milk contains less lactose than fresh milk, and that it also may contain bacterial lactase that may be activated when the fermented milk reaches the gut.

Galactosemia

Digestion of lactose in the intestine, and fermentation of milk gives increased concentrations of galactose. Galactose is catabolized by the Leloir pathway by phosphorylation at position 1, and then converted to UDP-galactose and glucose-1-phosphate. Defects in enzymes in this pathway may result in galactosemia in humans, and early onset cataract. In young women ovarian failure at a very early age has been observed following galactose accumulation. Cramer et al.studied the relation between age-specific fertility rates, the prevalence of adult hypolactasia and per capita milk consumption. They found that fertility at high ages is lower with high per capita consumption of milk and greater ability to digest its lactose component. These demographic data thus add to existing evidence that dietary galactose may deleteriously affect ovarian function.

The level of galactose in fermented milk products depends on growth conditions of the different organisms and fermentation time, and for example after 24 h fermentation, the concentration of galactose has been reported to about 20 g/litre . A study in rats showed that administration of galactose in the form of lactose seemed to be less toxic than when galactose was fed High levels of galactose as well as glucose may cause glycation of proteins, form advanced glycation end products, and the activation of polyol metabolism. This may accelerate generation of reactive oxygen species (ROS) and increases in oxidative chemical modification of lipids, DNA, and proteins in various tissues.

Possible Concerns of Milk in Current Use

Within modern societies the milk has to be treated in different ways to keep for several days. This processing includes steps that may be of concern. In fresh milk each lipid globule is surrounded by apical plasma membrane from the mammary epithelial cell. It is not known whether the milk homogenisation, when the fat globules with their globule membrane are broken up into many new small lipid droplets with just a small fragment of the originating membrane, might have health implications.

Proteins and peptides are heat sensitive, and their bioactivity may be reduced by pasteurisation of milk. Heating of milk may also result in the formation of potentially harmful new products i.e. when carbohydrates in milk react with proteins. Also the amount of some vitamins and antioxidants may be reduced by heating. Glutathione may easily be destroyed during storage The glutathione concentration in human breast milk was reduced by 81, 79 and 73% by storage at either -20° C, 4° C or at room temperature for 2 h, respectively To treat milk in a way that preserves the vitamins, proteins and peptides is therefore an important task and a challenge for the dairy industry. Some dairies now membrane filtrate the milk in stead of pasteurisation, and application of non-thermal processing technologies may give health benefits.

Several components in bovine milk which are of great importance in human nutrition may be significantly altered by the feeding regime The principal effects of feeding on milk content of these components are summarized and briefly discussed below.

Fat Content and Composition

The fatty acids of bovine milk are derived from two sources. The first source is fatty acids supplied to the udder by the blood, composed of fatty acids absorbed from the intestine and mobilized from the adipose fat tissue, mainly palmitic acid (16:0), stearic acid (18:0) and longer chained fatty acids. The second source is derived from circulating blood acetate and butyrate produced during fermentation in the rumen (de novo synthesis), and fatty acids up till 14 carbon atoms are synthesised in the udder. Palmitic acid in milk originates from both de novo synthesis and from circulating blood. Due to the extensively biohydrogenation of dietary unsaturated fatty acids in the rumen, the supply of these fatty acids to the udder is low. However, in the udder desaturation of fatty acids like 12:0, 14:0, 16:0 and 18:0 take place, and the products being 12:1, 14:1, 16:1 and 18:1, respectively. The preferred substrate for the desaturating enzyme; delta-9-desaturase, is stearic acid. Therefore is bovine milk a relatively good source of oleic acid (18:1, cis 9). The udder enzymes can not make double bonds in omega-3 and omega-6 positions. Consequently, milk content of linoleic acid and alpha-linolenic acid depends on the supply of these to the udder.

Conjugated linoleic acid (9c,11t-CLA) in milk originates from two sources. A small part originates from incomplete biohydrogenation of linoleic acid in the rumen which are absorbed from the small intestine transported to the udder and included in the fat synthesis. Most of the 9c,11t-CLA originates, however, from vaccenic acid which is an intermediate from biohydrogenation of unsaturated fatty acids in the rumen. After absorption and transportation by the blood to the udder, a portion of the vaccenic acid is desaturated by delta-9-desaturase to CLA. There is a close positive correlation between milk content of vaccenic acid and 9c,11t-CLA.

The effect of feed on milk fat content and fatty acid composition is comprehensively discussed There are large variations in fat synthesis in the udder, and the fat content and fatty acid composition are the most modifiable of the main components in milk. There are seasonal variations for the major fatty acids Milk from grazing dairy cows contain significantly higher proportion of oleic acid than milk produced on traditional indoor feeding composed of concentrates and conserved roughages Typically, CLA content in milk produced on pasture is at least twice of that obtained by indoor feeding Moreover, the proportion of alpha-linolenic acid increases more than linoleic acid, resulting in a lower ratio between omega-6 and omega-3 fatty acids. Milk fat from cows fed an in-door diet consisting of conserved grass and concentrate have a ratio between omega-6 and omega-3 fatty acids of about 4:1 but in summer when the cows are out on pasture and have a high intake of grass the ratio may be reduced to about 2:1 These positive effects of pasture on the fatty acid composition of milk are mainly attributed to the high content of polyunsaturated fatty acids, especially alpha-linolenic acid, in grasses at early stage of maturity. Components in bovine milk and their chances to be modified according to feeding strategies, substrates involved in their synthesis and feeding strategies that may be used.

Protein Content and Composition

In general, milk protein content is relatively unresponsive to feeding factors. However, under most conditions, energy-, but also protein supply, is feed related factors with most pronounced effect on milk protein content. Milk protein content may be negatively influenced by high intake of dietary fat by the lactating cow. Thus, there may be a conflict between milk fatty acid composition and protein content. The feeding regime has only small impact on the proportion of the different types of milk proteins and consequently the amino acid composition, and will therefore not be further discussed in this review. However, heat treatment of dairy products leads to structural changes of proteins and main whey proteins are modified to lactulosyl residues.

Content of Minerals

Milk concentrations of some minerals that are of special importance in human nutrition The calcium concentration in milk is relatively constant, with some variations throughout lactation. Most of the calcium is in the aqueous compartment and it is primarly (65%) associated with casein The calcium concentration in milk is relatively constant because milk content of casein is unresponsive to feeding factors. Milk content of magnesium and zinc also show only small variations.

The concentration of selenium in bovine milk is related to selenium concentration in the feed, and there are great variations worldwide. In South Dakota the selenium concentration in milk is reported to be between 160 and 1300 μg/l, whereas the concentrations in milk from low-selenium regions may be from 5 to 30 ug/l as in Scandinavia and northern Europe. A Swedish study showed that the average selenium concentration in milk was 14 ug/l and the concentration was more than doubled after supplementation with 3 mg selenium daily from selenium-enriched yeast.

As much as about 25 per cent of the iodine intake may be excreted in milk Therefore, milk content of iodine also varies depending on the iodine content and availability in the feeds used. A study on milk and milk products in Norway , showed that milk from the summer season had significantly lower iodine concentration (88 μg/l) compared with milk from the winter season (232 μg/l). This is explained by the use of more supplementary feeds enriched with iodine during the winter season. Dairy products supply much of the dietary intake of iodine; in Norway the most and in US the second most of the iodine intake.

Content of Vitamins

Vitamins are not synthesized in the udder. Milk content of the fat-soluble vitamins A and E reflects their content in the feed). In general, the content of these vitamins in feed plants decrease with maturity and are higher in fresh than conserved material. There are therefore regional and seasonal variations in these vitamins due to feeding regimes , with the highest concentrations in fresh grasses at early stage of maturity. For example a study in Finland shows that the concentration of vitamin E in milk is 3–4 times higher during summer than in winter Enriching the supplementary feeds with proper sources of these vitamins may increase the milk content during the winter season.

All vitamins of the B-complex (riboflavin and vitamin B_{12}, in are synthesized by the rumen microbes, normally in sufficient amounts to cover the animal's needs. Milk content of B-vitamins are, however, relatively unrelated to their intake because the amount synthesized by the rumen microbes are unregulated according to the amount ingested.

CONCLUSION

Consumption of 0.5 litre milk daily supplies a significant amount of many of the nutrients that are required daily. Milk components take part in metabolism in several ways; by providing essential amino acids, vitamins, minerals and fatty acids, or by affecting absorption of nutrients. Milk fat is diverse with a wide-ranging spectrum of fatty acids and lipids. Milk fat has been notified for decades, but as discussed in this review a moderate intake of milk fat has no negative health effects, on the contrary, many milk fat component have important roles in the body. Milk protein is especially rich in amino acids that stimulates muscle synthesis, and some proteins and peptides in milk have positive health effect e.g. on blood pressure, inflammation, oxidation and tissue development. Fermented milk has special health-promoting properties, e.g. stimulation of immune response and protection against cancer, virus and allergy, and fermented milk and full-fat milk may also delay gastric emptying from the stomach and possibly have an effect on appetite regulation. For some individuals, milk proteins, fat or milk sugar may cause health problems. Heat treatment of milk may also result in reduction of bioactive compounds and formation of potentially harmful products of carbohyd rates and proteins. Milk can be significantly altered by changing the feeding regimes. Content of several fatty acids such as c9, t11-CLA and the ratio between omega-6 and omega-3 fatty acids are affected by the amount of grass and supplemental feeds (concentrate) in the diet. Milk content of several vitamins and minerals are also influenced by the cow's diet. Iodine and selenium are examples of trace elements that may be added to the feed, and thereby milk can be a good food source of these elements.

INDEX

D